T0093932

Centenarians

Calogero Caruso

Editor

Centenarians

An Example of Positive Biology

 Springer

Editor
Calogero Caruso
Department of Biomedicine, Neurosciences and Advanced Diagnostics
University of Palermo
Palermo, Italy

ISBN 978-3-030-20761-8   ISBN 978-3-030-20762-5   (eBook)
https://doi.org/10.1007/978-3-030-20762-5

This Springer imprint is published by the registered company Springer Nature Switzerland AG
The registered company address is: Gewerbestrasse 11, 6330 Cham, Switzerland

Preface

People worldwide are living longer.

At global level, the share of 80+ people rose from 0.6% in 1950 (15 million) to around 1.6% (110 million) in 2011, and it is expected to reach 4% (400 million) by 2050. The global population is projected to be 3.7 times bigger in 2050 than in 1950, but the number of 60+ people will increase by 10%, while the number of 80+ people will increase by 26%. The increase in lifespan does not coincide with the increase in health-span, i.e. the period of life free from serious chronic diseases and disability.

Improving the quality of life of oldest is becoming a priority due to the continuous increase in the number of this population who is at risk of frailty. This makes the studies of the processes involved in longevity of great importance.

Most biomedical researches are "negative biology", because the study of the disease is its central heart, focusing on the causes of the diseases. On the contrary, a different approach is possible, called "positive biology". Instead of placing diseases at the centre of research, positive biology searches, for understanding the causes of positive phenotypes, to explain the biological mechanisms of health and well-being. This means understanding why some individuals, namely the centenarians, have escaped neonatal mortality, infectious diseases in the pre-antibiotic era and the fatal outcomes of age-related diseases, thus living more than 100 years. The knowledge born from this approach could allow modulating the ageing rate by providing valuable information on lifestyle to achieve healthy ageing.

Furthermore, the study of centenarians could provide important indications on how to build drugs that can slow or delay ageing, with benefits for those who are more vulnerable to disease and disability. The identification of the factors that predispose to a long and healthy life is therefore of enormous interest for translational medicine.

It is known that the longevity phenotype is the result of a positive combination between genetic, epigenetic, stochastic and lifestyle factors. So, the analysis of all the known parameters that can influence these single elements, or their interaction, can give new possible information to delineate a sort of longevity signature.

In the different chapters of this book (see in Contents), a detailed analysis of the mechanisms involved in achieving longevity is performed. The role of chance, genetics, epigenetics, sex/gender, education and socio-economic level, social support and stress management, diet and nutrition, microbiota and pathogen burden, physical activity, immuno-inflammatory responses and oxidative stress are depicted.

Palermo, Italy   Calogero Caruso

Contents

Chance and Causality in Ageing and Longevity

Giulia Accardi, Anna Aiello, Sonya Vasto,
and Calogero Caruso

1.1 Introduction

Longevity is not a matter of genes.

This is the message that appeared last year in all the newspapers of the world, according to a study due to a joint venture between the statisticians of Ancestry and Calico Life Sciences that has dissected the genealogical trees of 400 million individuals, tracing back generations, and including dates of birth, death, places, and family ties. The genes would have little to do with longevity: in a percentage perhaps even less than 10% [1]. However, this extensive study has analysed the influence of genetics in terms of lifespan, but not in terms of longevity.

Longevity may be defined in relative and absolute terms [2]. Longevity, indeed, may be considered a concept country/population specific, since different populations/countries show great variability of their life expectancy, represented by the age reached by 50% of a given population, owing to historical, anthropological, and socio-economic differences. In "absolute" terms, instead, longevity is defined according to the maximum lifespan attained and scientifically validated by human beings on the Earth (Chap. 4). The threshold of exceptional longevity is regarded the canonical age of 100.

Giulia Accardi and Anna Aiello contributed equally with all other contributors.

G. Accardi · A. Aiello · C. Caruso (✉)
Laboratory of Immunopathology and Immunosenescence, Department of Biomedicine, Neurosciences and Advanced Diagnostics, University of Palermo, Palermo, Italy
e-mail: giulia.accardi@unipa.it; anna.aiello@unipa.it; calogero.caruso@unipa.it

S. Vasto
Department of Biological, Chemical and Pharmaceutical Sciences and Technologies, University of Palermo, Palermo, Italy
e-mail: sonya.vasto@unipa.it

Demographic evidences revealed a continuing decrease in the mortality rate in old age and an increase in the maximum age at death, which could gradually extend human longevity. Human longevity, therefore, might not be subject to strict constraints. However, analysing the global demographic data, Dong et al. [3] have shown that improvements in survival with age tend to decrease after age 100 and that the age at death of the oldest person in the world has not increased since the 1990s (Chap. 4). In a recent study, analysing data on all Italians aged 105 and older between 2009 and 2015 (born in 1896–1910), Barbi et al. [4] have provided the best evidence to date for the existence of extreme-age mortality plateaus. These studies suggest that the maximum duration of human life is fixed and subject to natural constraints.

There is no doubt that human beings are the result of their genes, but they would not be the same without the experiences that they have had since in the mother's womb. As an example, the month of birth, a proxy for early-life environmental influences (i.e. epigenetics), affects the possibility of achieving age 100 [5]. On the other hand, each individual would not be the same if he was born in another part of the world.

Both immunological repertoire and the architecture of brain, which influence survival, are also linked to many small causal events. These are examples of the effect of chance [6]. Chance is just that the random occurrence, i.e. an event happening not according to a plan, that surrounds everybody. Chance is the product of entropy, i.e. the natural tendency of the universe to move towards disorder. It is a major player in all kinds of life, animal, and vegetal. Of course, it is a matter of choice as everyone manages the various things that chance throws at everybody. It is sometimes possible to modify the probabilities of a random event. As an example, if individuals never buy a lottery ticket, then it is not possible for them to win a lottery. If they buy a ticket, there is a chance they could (https://www.enotes.com/homework-help/what-role-does-fate-chance-play-ones-life-407539).

In biology, chance can be defined as the occurrence of events in the absence of any obvious intentional cause. It must be distinguished from life circumstances, events, or facts that cause or help to cause something to happen, e.g. the death in war of a potential centenarian [6].

To ask if longevity depends on the environment or genetics is legitimate but oversimplified. Answering means to consider everything is important, luck (chance and life circumstances), lifestyle (nutrition, physical activity, and environmental exposures, all also affecting epigenetics), life experiences (pathogen burden, stress management and social support, education), and biology, with sex and DNA (genetics and epigenetics) in the front row [6–8]. Some of these are effects of a cause.

Causality is what connects one process, the cause, with another process or state, the effect, where the cause is completely or partly responsible for the effect, and the effect is completely or partly dependent on the cause. In biology, a process has many causes, the causal factors for it, and all lie in the past. In turn, an effect can be a causal factor for many other effects (https://en.wikipedia.org/wiki/ Causality). To establish functional causality, three criteria must be met: (1) a strong association between the two factors studied must be demonstrated; (2) there must be a plausible mechanism to explain the cause and effect relationship; (3) other obvious competing explanations about the observed association should be excluded [9].

The exhausting research of a way to achieve longevity and well-being has always been a constant in human history. Nowadays, reaching the age of 80+ is a common goal, but this was a privilege up to 100 years ago [10]. Throughout history, the path leading to longevity was full of different obstacles but always marked by causality and chance.

In this chapter, the role of chance and causality in achieving longevity is discussed. The genetic and epigenetic causes of longevity are described in Chaps. 6 and 7, respectively. Some aspects of lifestyle involved in longevity, i.e. diet and nutrition are illustrated in Chaps. 5, 10, and 11, whereas in this chapter the role of nutrient-sensing pathways (NSPs) and their modulation are discussed. The role of physical activity in attainment of longevity is mentioned in Chap. 4 and examined in depth in this chapter. Some phenotypic aspects including stress management and social support are defined in Chaps. 2, 3, and 8 and further analysed in this chapter. Lastly, in this chapter are discussed the roles played by education, sex, and immune-inflammatory responses, encompassing oxidative stress to environmental agents, including pathogen load (the role of microbiota is studied in Chap. 9).

In this book, indeed, studies on different aspects of centenarians are extensively reported. Centenarians show relatively good health, being able to perform their routine daily life and to escape or delay age-related diseases. As discussed by Caruso et al. [7], the aim is to understand, through a "positive biology" approach, how to prevent, reduce, or delay elderly frailty and disability. Rather than making diseases the central focus of studies, "positive biology" seeks to understand the causes of positive phenotypes, trying to explain the biological mechanisms of health and well-being [11].

1.2 Role of Chance in Ageing and Longevity

Several centuries ago, Democritus said that everything that exists in the universe is the result of chance and necessity (http://www.normalesup.org/~adanchin/causeries/Atomists.html#N1). Then, Epicurus tried to visualize the chance by stating that, from time to time, the normally straight paths of the atoms of the universe bend slightly and the atoms wander. If we consider mispairing during DNA replication, perhaps it was not very far from what happens [12].

As discussed by Monod [13], the alterations in DNA are accidental and happen at random. Since they represent the only possible source of modification of the genetic text, which in turn is the sole depository of the hereditary structures of the organism, it necessarily follows that the only chance is the origin of every novelty, of every creation both in the biosphere and in the soma.

In fact, living beings are characterized by the ability to transmit their genetic structure to subsequent generations. From the moment in which the change in the DNA structure has occurred, it will be automatically and faithfully replicated and translated. The same occurs in the soma, from stem cells to somatic cells. Out of the realm of pure chance, it enters that of necessity, of the most inexorable determinations.

Random processes play a role in physiological and pathological events, alongside genetics, epigenetics, lifestyle, and environment. Intrinsic randomness in biological occurrences contributes to the individuality of living organisms, since it influences phenotypic variability. They are subject to nature laws and genetic programs where both Brownian random motion, i.e. the disordered motion of particles suspended in fluids resulting from their collision with the fast-moving water molecules in the fluid, and crossing over contribute to leaving space for the chance [14].

There is increasing evidence of the intrinsic random nature of gene expression and macromolecular biosynthesis, since many genes are transcribed to minimal amounts of mRNA per cell, which can cause large fluctuations in biosynthesis. Genomic instability is another important source of intrinsic variability, as shown in aged mice, which have a mutation frequency up to 10^4 per gene per cell. Somatic mutations are random events since they result from mispairing, originating from the equilibrium that exists in solution between tautomeric forms of the purine and pyrimidine bases. Epimutations may also occur through random changes of DNA methylation patterns, affecting gene expression [14, 15].

Therefore, genetically identical cells exposed to the same environmental conditions can show significant variation in molecular content and marked differences in phenotypic characteristics. This variability can provide the flexibility needed by cells to adapt to fluctuating environments or respond to sudden stresses. It is a mechanism by which population heterogeneity can be established during cellular differentiation and development. How cells adopt a particular fate is usually thought of as being deterministic, and in the large majority of cases, it is. In some cases, however, cells choose one or another pathway of differentiation randomly, without apparent regard to environment or history. However, certain developmental decisions are imposed on systems otherwise intrinsically random. Nature knows how to make deterministic decisions, but, in contrast to Einstein's view of the universe, nature also knows how to leave certain decisions to a roll of the dice when it is to her advantage [16, 17].

The role of chance in ageing and longevity is clearly demonstrated by experimental studies conducted on inbred mice. They have the same genome as well as the same housing conditions, but show different lifespan, up to 50% higher, contrary to expectations. This proves that in living organisms there is a component of continuous random microvariations, due to the events reported above. These, accumulating over time, amplify the differences between individuals, manifesting in a striking way in older ages [15].

1.3 Socio-economic Development and Life Expectancy: Causal Role of Education

In the Western world, environmental improvements beginning in the 1900s are thought to have extended the average lifespan dramatically, and, later, the number of centenarians, due to significantly greater availability of food and clean water, better housing and living conditions, reduced exposure to infectious diseases, and access to medical care (https://ghr.nlm.nih.gov/primer/traits/longevity). Health and life expectancy are considered, in fact, closely associated with socio-economic development. Better nutrition and greater availability of health care associated with higher income have been widely regarded as primary determinants of decreasing historical and contemporary mortality. In many of the studies of this issue, the assumption that income is the most important driver of mortality decline has been an unquestioned starting point.

A very different picture was drawn by other studies that pointed out the role of education. Thereafter, the question of the causal nature of the effects of education and income on health has attracted much controversy.

Recently, Lutz and Kebede [9] have addressed the hypothesis that the apparent statistical association between income and health could in fact be a largely spurious association resulting from the fact that improving educational attainment is a key determinant of both better health and rising incomes.

To this end, they verified both theories by analysing data on 174 countries, collected between 1970 and 2015. They compared information on life expectancy with those on education and per capita income and noted that although there was a correlation between the life expectancy and both factors i.e. higher salary and better education, the relationship between longer life and education was more linear and constant. A more complete education is a good predictive factor of a longer existence, and this, scientists assume, is because a more solid culture leads to healthier choices of life, e.g. in nutrition or disease prevention. The correlation between wealth and longevity, that exists, even if less clear, is perhaps caused by the fact that a better education also gives access to higher and better-paid jobs, and therefore to greater possibilities of medical care and better nutrition. However, money would not be the ultimate cause of long-living people.

1.4 Role of Sex and Gender in Ageing and Longevity

In Western countries, women live 5–6 years more than men do. Furthermore, 85% of women are over 100 years old (Table 1.1). Although this gender difference has been accepted as normal, it is a relatively recent demographic phenomenon due to the reduction of infections and to the increase in the adult mortality rate attributed to cancer and cardiovascular disease. In particular, heart disease is the main condition associated with the increase in excess male mortality, making the strongest contributions in the birth cohorts of 1900–1935 [18].

Table 1.1 List of the ten countries with the highest life expectancy (https://en.m.wikipedia.org/wiki/List_of_countries_by_life_expectancy)

Men		Women	
Country	Life expectancy (years)	Country	Life expectancy (years)
Iceland	81.2	Japan	87.0
Switzerland	80.7	Spain	85.1
Australia	80.5	Switzerland	85.1
Israel	80.2	Singapore	85.1
Singapore	80.2	Italy	85.0
New Zealand	80.2	France	84.9
Italy	80.2	Australia	84.6
Japan	80.0	Korea	84.6
Sweden	80.0	Luxembourg	84.1
Luxembourg	79.7	Portugal	84.0

World life expectancy is 76.0

It is debated whether women live longer than men do for reasons of gender or sex, i.e. for cultural or biological differences. However, given that the females live longer than males in other animal species, this phenomenon cannot be attributed solely to cultural behaviour, but must be due, at least in part, to biological causes, such as the different production of hormones or the presence of the two X chromosomes. As an example, testosterone decreases levels of high density lipoproteins and increases levels of low density ones (LDL), while oestrogens have an opposite effect with beneficial effects on the risk of cardiovascular disease. Another possibility is that women, due to the menstrual cycle, have a relative lack of iron compared to men for about 30–40 years. Iron is a crucial catalyst for the mitochondrial production of free radicals as an effect of metabolism. Probably a reduction in available iron leads to a lower production of free radicals. Diets rich in iron have, in fact, been associated with a considerably greater risk of heart disease. A relative iron deficiency protects against bacterial infections. On the other hand, oestrogens, by interacting with their receptors, mediate cytoprotective effects against reactive oxygen species (ROS), enhancing the levels of different mitochondrial antioxidant enzymes. Mitochondria of the cells of female individuals produce less ROS than those of the cells of male individuals. It follows that women compared with men are suffering from cardiovascular diseases such as heart attack or stroke, about 10 years later [19–21].

In addition, the control of immune-inflammatory responses plays a key role in successful ageing, and women have a more powerful immune response and better control of inflammatory responses [21, 22].

Testosterone, on the other hand, can lead to aggressive behaviour and a lifestyle at risk such as dangerous driving, alcohol abuse, and use of weapons: all behaviours that can increase the mortality rate. Not only biology, therefore, but also social factors contribute to longevity. Women tend to perform quieter activities within the family and with less professional activities and lifestyle for their health: they smoke less, they eat less, they cope better with stress, and they take better care of their body and their health. All these are gender differences, socially constructed sex, i.e. the characteristics that society outlines as masculine or feminine. As such, they are valid in today's older generations, but not in the future ones, since today's girls tend to copy many deleterious aspects of male culture (aggression, alcoholism, and smoking) [21].

1.5   Inflamm-ageing and Oxidative Stress as Causal Factors of Ageing

As defined, ageing is characterized by the breakdown of a self-organizing system and the reduced ability to adapt to the environment [2]. This goes hand in hand with the growth of a low level of chronic inflammation state noted as "inflamm-ageing" [23].

Most explanations of the increase in life expectancy at older ages over history emphasize the importance of medical and public health factors of a particular historical period. However, as discussed in Sect. 1.3, according to Lutz and Kebede [9], the role of education in this dramatic increase of life expectancy seems to be prominent.

On the other hand, the reduction in lifetime exposure to infectious diseases and other sources of inflammation has also made an important contribution to the historical decline in old-age mortality. Analysis of birth cohorts across the lifespan since 1751 in Sweden reveals strong associations between early-age mortality and subsequent mortality in the same cohorts. Decreased inflammation during early life has led directly to a decrease in morbidity and mortality resulting from chronic conditions in old age. This hypothesis is supported by researches linking individual exposure to past infection, i.e. to pathogen burden, to levels of chronic inflammation, and to increased risk of heart attack, stroke, and cancer [24].

The most important link between the pathogen burden and inflammation is represented by Toll-like receptor (TLR) proteins. These proteins play key roles in immune responses against infections. In general, TLRs can recognize structurally conserved pathogen-associated molecular pattern molecules (PAMPs), which are derived from invading bacteria or virus. Recognition of PAMPs by TLR-expressing cells results in acute responses necessary to clear the pathogens in the host. However, TLRs are also involved in the pathogenesis of autoimmune, chronic inflammatory, and infectious diseases. Recently, evidence linking TLR4 with the pathophysiology of age-related chronic diseases (i.e. cardiovascular diseases, neurodegenerative diseases, cancer) has been presented. It is noteworthy that inflammation is the common and pivotal background shared by these pathological conditions. Evidence discussed in that paper indicates the involvement of genetic TLR4 polymorphisms in the susceptibility to those age-related diseases, since TLR4-mediated inflammation from bacterial and viral infection or other endogenous molecules can influence the development of these diseases. Given the complex inheritance patterns of these polygenetic traits, the impact of any single allele on the trait under study is obviously modest [25].

Inflamm-ageing is an important risk factor for both morbidity and mortality in the elderly that makes them more vulnerable to above reported age-associated diseases. Inflammation is not per se a negative phenomenon. In fact, in response to cell injury elicited by trauma or infection the inflammatory response sets in, constituting a complex network of molecular and cellular interactions directed to facilitate a return to physiological homeostasis and tissue repair. If tissue health is not restored or the inflammatory trigger is not cleared, acute inflammation may become a chronic condition that continuously damages the surrounding tissues. The collateral damage caused by this type of inflammation usually accumulates slowly, sometimes asymptomatically for years but can eventually lead to severe tissue deterioration [26]. Indeed, according to the theory of antagonistic pleiotropy, in early life, when natural selection is strong, the inflammation should be beneficial. It becomes dangerous during ageing when selection is weak [27].

As reported in the Encyclopedia of Immunobiology [22], in aged people, several changes of both innate and acquired immunity have been described and viewed as deleterious, hence the term immunosenescence (Chap. 3). It is linked not only to the functional decline associated with the passage of time but also to antigen burden to which an individual has been exposed during lifetime. The elderly immune system is characterized by continuous reshaping and shrinkage of the immune repertoire by persistent antigenic challenges. These changes lead to a poor response to newly encountered microbial antigens, including vaccines, as well as to a shift of the immune system towards an inflammatory, autoimmune, Th2 profile (i.e. humoral immune response) [23].

The long-life chronic antigenic stress, determining a progressive activation of macrophages, contributes to the inflammatory status observed in the elderly. This chronic state of low-grade inflammation, inflamm-ageing, is characterized by increased levels of pro-inflammatory cytokines and acute phase proteins, which increase is a worse prognostic factor for all causes of death. This inflamm-ageing state is implicated in the pathogenesis of age-related inflammatory diseases [23].

In a large study performed by Arai et al. [28], centenarians with the lowest levels of markers of chronic inflammation were able to maintain good cognition and independence for the longest periods. Those with less inflammation also experienced the greatest longevity. Inflammation was a strong predictor of cognitive capacity in semi-supercentenarians (>105 years). The offspring of centenarians tended to have lower markers of chronic inflammation, meaning your chances of living a long, healthy life are to some extent familial (Chap. 3). This indicates that someone who will (probably) become a centenarian is able to keep inflammation down for longer. In addition, Storci et al. [29] have reported on the anti-inflammatory molecular make-up of centenarians' fibroblasts, low levels of interleukin (IL)-6, type 1 interferon β, and pro-inflammatory microRNAs.

Oxidative stress plays an important role in determining and maintaining the observed low-grade inflammation, which, in turn, contributes to oxidative stress [30, 31].

The use of oxygen by the mitochondria of the aerobic cells generates potentially deleterious reactive oxygen metabolites, which increase with ageing and are negatively associated with life expectancy of organisms [32]. The hypothesis that age-associated functional losses are also due to the accumulation of ROS and nitrogen species gave rise to the oxidative stress theory of ageing. This postulates that a slow accumulation of oxidative damage to macromolecules like DNA, proteins, and lipids causes age-associated reductions in physiologic functions [33].

The oxidative stress is determined by an imbalance between pro-oxidant and anti-oxidant factors that gets out of control in favour of the pro-oxidants. It results in macromolecular damage and is considered one of the major causal factors of cellular senescence [34], as well as telomere shortening [35]. The levels of the enzymes involved in the clearance of the free radicals in the cytosol, like superoxide dismutase, catalase, and glutathione peroxidase, are decreased in ageing cells and contribute to the increased cellular oxidative stress. Moreover, manganese superoxide dismutase, an antioxidant enzyme located in the mitochondria, which protects macrophages from apoptosis induced by oxidized LDL, is also decreased in ageing [36].

The oxidative damage produced by free radicals promotes the inflammatory state by two major pathways: the TLRs and the inflammasome [37, 38].

In this case, TLRs seem to be activated by damage-associated molecular pattern molecules released during conditions of oxidative stress [37].

Oxidative stress can induce an inflammatory response also through inflammasome activation. The inflammasome is a cytosolic multiprotein complex that cleaves pro-IL-1 and pro-IL-18 into active cytokines, immediately after the assembly with caspase 1. Several inflammasome complexes exist and, among them, the Nalp3 inflammasome has been recently shown to be directly activated by the presence of sustained amounts of ROS [39].

The oxidative burst associated with immune-inflammatory responses upregulates the formation of ROS and the overall oxidative stress response. Overproduced free radicals will create a circular loop of TLR and Nalp3 activation [39].

Therefore, ageing is associated with an increase in plasma concentrations of fasting free radicals. It is due to a raise in pro-oxidant factors and to a decline in antioxidant mechanisms [40]. However, this does not occur in the best model of successful ageing, the healthy centenarians. In several centenarian groups, from different geographical areas, some indices of oxidative stress (i.e. lipid peroxides, oxidized/reduced glutathione ratio) were lower than in aged subjects [41]. In a study with 32 healthy centenarians, they were characterized by the highest levels of vitamins A and E, well known as antioxidant vitamins, whereas the activities of both plasma and red blood cell superoxide dismutase, which increased with age, decreased in centenarians. It is clear that healthy centenarians show a particular profile in which high levels of vitamin A and vitamin E seem to be important in their extreme longevity [42]. From a study conducted on 153 centenarians from different geographic areas, it again emerged that the plasma concentration of vitamins E and A resulted greater in centenarians than in aged subjects. However, the centenarians from Sardinia had vitamin A and E levels not significantly different from those of younger subjects of the same geographical area. Thus, in Sardinian centenarians probably other factors play a more important role [43].

In addition to immunosenescence, pathogen burden, and oxidative stress, a wide range of different stimuli contribute to provoking and/or modulating inflamm-ageing (Sect. 1.7, Chaps. 5, 9, and 11) [23, 44–47].

However, from discussed data, it is clear that the control of inflammation and oxidative stress may enhance the possibility of becoming centenarians.

1.6   The (De)regulation of Nutrient-Sensing Pathways in Ageing and Longevity

Healthy dietary habits and physical activity (Sect. 1.7) are the most important modi-fiable factors that can affect the achievement of a healthy ageing phenotype [48].

Since many years, interventions that try to slow the rate of ageing and increase lifespan have interested many scientists [49]. With the aim to develop preventive and therapeutic measures, it was important to start managing the daily habits. Nutrition is, probably, the most important one. Indeed, healthy nutrition plays a significant role in delaying and reducing the risk of developing age-related diseases, and represents an integrative and tangible approach [46].

Several studies have suggested that both in animal models and humans, dietary intervention can prevent or decrease various age-related diseases, by positively regulating some ageing processes through modulation of the nutrient-sensing path-ways (NSPs). The NSPs are a signalling cascade activated by the level of nutrients, such as carbohydrates or protein (or amino acids), that trigger signals. This results in a downstream activation of genes involved in ageing process. The more represen-tative links to longevity, both in humans and in model organisms, are the insulin/insulin-like growth factor-1 (IGF-1), the mammalian target of rapamycin (mTOR), and the sirtuin (SIRT) pathways. From yeast to human, they are genetically con-served although the molecular complexity varies among them (Fig. 1.1) [46].

The insulin/IGF-1 signalling cascade starts from the binding of insulin or IGF-1 to the insulin/IGF-1 receptor (IGF-1R) that triggers many molecular events, such as the activation of the phosphoinositide-3-kinase. It leads to the activation of protein-kinase B that can stimulate the nuclear factor kappa-light-chain-enhancer of acti-vated B cells (NF-κB) signalling, involved in the modulation of immune-inflammatory responses [50, 51]. The translocation of NF-κB to the nucleus and its binding to the DNA provoke the transcription of a number of genes, including pro-inflammatory mediators, but also anti-inflammatory cytokines [46, 51, 52].

The modulation of this pathway, possibly due to a low glycaemic and protein intake and, also maybe, to the action of some antioxidant and anti-inflammatory molecules, could lower insulin/IGF-1 signal and stimulate the action of different transcription factors (TFs) [46, 53].

Glucose and growth hormone (GH) influence insulin and IGF-1 levels, respec-tively. During fasting, both GH and insulin decrease with a consequent reduction in IGF-1 circulating levels as shown in several model organisms [54]. Paradoxically, GH and IGF-1 levels decrease during normal ageing [55]. It is a common char-acteristic of both physiological and accelerated ageing, while a constitutive

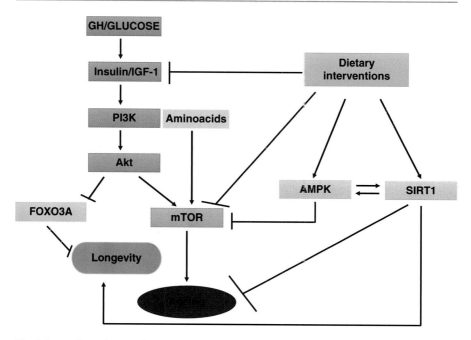

Fig. 1.1 A schematic overview of nutrient-sensing pathways and their effects on ageing and longevity (see the text for explanation). Abbreviations: *GH* growth hormone, *IGF-1* insulin growth factor-1, *PI3K* phosphatidylinositol 3-kinase, *Akt* protein kinase B, *AMPK* AMP-activated protein kinase C, *SIRT1* sirtuin 1, *mTOR* mammalian target of rapamycin, *FOXO3A* Forkhead box O 3A

downregulation of insulin/IGF-1 pathway extends longevity. These apparently contradictory observations could be explained by a model according to which the down-modulation of insulin/IGF-1 reflects a teleonomic response to minimize cell growth and metabolism in the context of systemic damage [56].

Forkhead box O 3A (FOXO3A), extensively studied for its role in longevity (Chap. 6), is one of the TFs upregulated as a result of downregulation of the insulin/IGF-1 pathway. It is involved in the transcription of molecules that take part in cellular homeostasis. The gene encoding this protein belongs to the FOXO family genes that have a typical DNA-binding forkhead box domain. It is one of the orthologues of DAF-16 in *Caenorhabditis elegans*, a TF involved in stress resistance and longevity [46, 57, 58].

FOXO3A interacts with sirtuins (SIRTs), considered anti-ageing molecules in model organisms. The sirtuins pathway is another related and interconnected nutrient-sensing system. They act in the opposite direction to insulin/IGF-1, indeed their activation signals are scarcity and catabolism, by detecting high NAD$^+$ levels [59]. Accordingly, their upregulation favours healthy ageing. The sirtuins belong to a well-known family of nicotine adenine dinucleotide (NAD+)-dependent enzymes, initially described as transcription-silencing histone deacetylases in yeast. In mammals, there are seven yeast homolog sirtuins (SIRT1–SIRT7) that exhibit differentially subcellular localization, substrate affinity, and activity. In particular, SIRT1 deacetylates FOXO3A, modulating its response to oxidative stress [60]. It also has a protective role in endothelial function, preventing cardiovascular diseases. Similarly, SIRT3 is implicated in metabolism and mitochondrial function, including ROS detoxification. On the contrary, the overexpression of SIRT7 is observed mainly in cancer cells, while its depletion contributes to the prevention of ageing. SIRT6 has been identified as a critical regulator of transcription, genome stability, telomere integrity, DNA repair, and metabolic homeostasis, with an important effect on the ageing process. SIRT2, especially active in the cytoplasm, has anti-inflammatory effects, inactivating NF-κB [61].

Downstream of IGF-1R there is mTOR pathway, composed of mTOR complex 1 (mTORC1) and mTOR complex 2. mTORC1 is activated by high amino acid concentrations, and insulin and IGF-1 levels. It regulates growth, metabolism, and stress response through the regulation of transcription, translation, and autophagy, a cytoprotective process [62]. This is a self-degradative process, important for the balance of sources of energy in development and in the response to nutrient stress. It, in case of reduction, promotes inflamm-ageing and unhealthy ageing [63]. Its inhibition by the 5′ AMP-activated protein kinase, a key sensor of the cellular energy state, activated by low levels of ATP, causes stress resistance and reduced age-related inflammatory status by the NF-κB pathway [62].

Thus, intense anabolic activity, signalled through the insulin/IGF-1 or the mTORC1 pathways, accelerates ageing and increases the risk of morbidity and mortality. In fact, the activation of these metabolic pathways is characterized by inflammation and mitochondrial dysfunction, with an increase of oxidative stress and a reduction of autophagy. Through the downregulation of IGF-1 and mTOR cascade or the upregulation of sirtuins, it is possible to extend lifespan in various model organisms, including mammals. These effects are also obtained by the presence of specific single-nucleotide polymorphisms in genes encoding proteins involved in NSPs, such as IGF-1 receptor (IGF-1R) and FOXO3A [58, 64, 65].

In model organisms and humans, the possibility to modulate ageing, with a non-invasive method, has been identified in some dietary interventions. Although the complex relationship between nutrition, the ageing process, and healthy ageing is not completely understood in humans, specific dietary changes, such as dietary restriction, the reduction of glycaemic and protein intake or the elimination of trans and saturated fats, the increased intake of omega-3, vitamins, micronutrients, and antioxidants, can help to minimize the inflamm-ageing. Similarly, appropriate intake of specific foods, the so-called functional foods or nutraceuticals, may confer health benefits, influencing the maintenance of immune homeostasis, and contributing, directly, to the reduction of inflammation and metabolic disorders [46, 66].

The crucial role of NSPs in attainment of longevity is clearly demonstrated by genetics and model animals.

There are many possible candidate genes for human longevity; however, of the many genes tested, only APOE and FOXO3 survived to association in independent populations. Several studies have noted specific FOXO3 SNPs associated with human longevity, in particular with FOXO3A rs2802292 G-allele (G>T). It is conceivable to speculate that hyper- or hypo-activation of this signalling pathway, due to genetic mutations that under- or overexpress regulative molecules, lead to different expression of homeostatic genes [64, 65].

As discussed in the Encyclopedia of Gerontology and Population Aging [67], the dietary restriction, defined as "dietary regimen in which specific food groups or micronutrients are reduced or removed from the diet", was first shown about 80 years ago to extend lifespan in rats [67, 68]. Caloric restriction (CR), the reduction of total calories intake by 20–40% without malnutrition, is the most well-known defined dietary intervention to delay ageing in model organisms. From yeast to primates, the effects of CR on health-span have been confirmed, suggesting a highly conserved role of some common ageing pathways [67, 69, 70]. In all models, the deprivation of specific nutrients is associated with the downregulation of NSPs [67, 71] and increased lifespan. A controlled randomized study on non-obese individuals reported that a 2-year 25% CR is feasible for humans and provides health benefits, such as decrease in inflammatory markers and cardio-metabolic risk factors [67, 72]. However, CR and other dietary interventions are not always applicable. Alternative approaches may be a close adherence to the Mediterranean or the Asiatic diet, which also includes healthy lifestyle. These diets reduce the risk of ageing-related diseases, promoting healthy ageing and longevity, delivering refined carbohydrates and amino acids in a pro-ageing way, and reducing NSPs, extending healthy lifespan [67].

1.7 Physical Activity and Longevity

Interaction with the environment remains crucial in the attainment of longevity as evidenced by the inverse relationship between mortality and a good level of physical activity, also at old age [73].

Being physically active, that does not mean to practise sport hard, but a lower and constant level of fitness seems to be associated with a higher life expectancy. Many studies evidence a reduction of mortality and a favourable effect on risk factors for mortality such as hypertension, high blood glucose and lipid levels, or hyperinsulinaemia in physically active people compared with that in inactive persons [74, 75]. A systematic review reported an increase of life expectancy by 0.4–6.9 years in people physically active, analysing the data extracted by 13 studies describing eight different cohorts [76].

The effect of constant and moderate exercise is observed both at systemic and cellular levels. For a comprehensive review of the effect of physical activity on systemic aspect of ageing, see Table 1.2.

At the cellular level, all the nine hallmarks of ageing [77] are influenced [78, 79]. Indeed, as proven by studies on model organisms, exercise confers genomic stability, favouring DNA repair, and NF-kB and peroxisome proliferator-activated receptor gamma coactivator 1-alpha signalling as well as reducing the number of DNA adducts [80–82]. Physical activity reduces the level of oxidative stress, and this could be an explanation of the known reduced telomere attrition in active people [83]. It can modulate DNA methylation, hence activating genes that encode

Table 1.2 The table reviews the main effects of regular physical activity in the elderly

System	Effect
Nervous	Increase of neurogenesis
	Reduction of neurodegeneration and cognitive alterations
Cardiovascular	Reduction of blood pressure
	Increase of cardiovascular functions (maximal cardiac output, regional blood flow and blood volume, body fluid regulation, endothelial and autonomic function, vagal tone and heart rate variability, and cardiac preconditioning)
Respiratory	Improvement of respiratory function by increasing ventilation and gas exchange
Metabolic	Raise of the resting metabolic rate, muscle protein synthesis, and fat oxidation
Musculoskeletal	Increase of muscle function and body composition by improving muscle strength and endurance, maintaining or regaining balance, motor control, and joint mobility
	Decrease of adiposity
	Increase of muscle mass
	Increase of bone density

Reference in [78]

telomere-stabilizing proteins and telomerase [84]. In addition, the methylation status of insulin signalling pathway genes in skeletal muscle after aerobic exercise seems to be different as well as that of genes involved in pro-inflammatory mechanisms, such as the caspase gene. This can contribute to modulation of age-related inflammatory status by physical activity [85, 86].

Physical exercise is also able to reduce loss of proteostasis, inducing autophagy, hence avoiding or decreasing the accumulation of protein debris [87]. Well known is the effect on glucose-sensing somatotrophic axis but evidences have demonstrated also an effect on all nutrient-sensing pathways. In particular, in this context, the modulation of this signalling is inverse to the one increasing longevity through dietary restriction. Indeed, the activation of these pathways by resistance exercise seems to be associated to the prevention of age-related sarcopenia by muscle protein synthesis [88]. Also, the three integrative hallmarks, cellular senescence, stem cell exhaustion, and altered intercellular communication, culprits of ageing phenotype, are influenced by physical exercise, as proven by human and model organism studies although it is not clear which are the molecular events directly correlated [78, 79].

Concerning the role anti-inflammatory of physical exercise, the association among exercise and pro- and anti-inflammatory responses is supported by many studies, although the limit between local and systemic inflammation, between protection and damage, is not well defined. The release of IL-6 has been demonstrated and correlates with stress hormone response, in particular in endurance exercise. Furthermore, IL-6 leads to neutrophil activation and trigger of anti-inflammatory mediators, such as IL-1ra and IL-10. Probably, the type of effect depends on the intensity of exercise, other than on the intake of energy and on the systemic status of subjects [89].

All these data have been confirmed highlighting that a period of physical inactivity is sufficient to significantly increase mortality risk [90].

In this context, the data are not always exhaustive to establish a cause/effect relationship. For instance, physical activity decreases the risk of being overweight/obese, reducing therefore the risk factors for chronic diseases. However, it is equally plausible that being overweight/obese limits the ability to engage in physical activity in the first place [91].

Actually, most of the data show the health-related effects conferred by physical activity, such as the improvement of physical function. However, no direct association, so the existence of specific claim, exists with lower risk of death in healthy individuals compared with control group by randomized controlled trial—the only one type of study that can prove the effective causal association [92].

1.8   Psychological Stress and Longevity

The interest related to the link between psychological aspect and longevity is relatively new and markedly interesting, although very difficult to measure. Scientific explanations exist about psychological implications in longevity although the clear molecular explanation is not understood. It seems that meditation positively affects some cognitive functions, such as attention and memory, stimulating brain function [93, 94].

Meditation is, therefore, one of the activities involving mind and body, with interesting consequences on lifespan. Actually, the use of this term is often erroneous if we consider the original meaning described in the ancient Vedic texts. However, in this context it refers to general practices involving senses, mind, intellect, and emotions, such as contemplation and concentration [95].

It is not only a news that inflammation and stress are strictly related, but it is also well known that the pathophysiology of age-related diseases has a proven inflammatory base. Meditation helps muscles and nerves relax, so, theoretically, it reduces the level of stress. Thanks to the growing body of evidence, we have scientific explanation about the association between this activity, regularly performed, and a slow rate of ageing. A study conducted in 2016 on a group of subjects performing long-term meditation, comparing with a group of healthy, age- and sex-matched, no-meditators, demonstrated less local stress (by reduction of flare and so neurogenic inflammation) in response to topic application of an irritating agent (capsaicin cream) and a better response to externally induced psychological stress, measuring lower salivary cortisol levels [96].

Life stress, both perceived stress and chronicity of stress, is also an accelerator of telomere shortening, one of the nine hallmarks of ageing [77, 97] (Chap. 8). The reduction of both length of telomeres and activity of telomerase is associated with age-related disease and shorter lifespan, so it is detrimental for longevity, especially at advanced age [97, 98]. Unexpectedly, an increase in telomerase activity due to chronic life stress was also demonstrated [99]. This dual response was defined "telomerase responsivity" [97] and could be explained as dose-dependent if we considered the intensity and the duration of stressors. Indeed, the sudden increase in telomerase activity after an acute stress response that implies the production of a huge of cortisol and oxidative stress molecules could be explained as a sort of protective effect. This was hypothesized highlighting the remarkable increase in telomerase activity in lymphocytes of mice stimulated with an antigen or a virus and the maintenance of a high activity in memory cells [100]. Therefore, it can be speculated that this is a sort of hormetic response. Hormesis is defined as "an adaptive response characterized by biphasic dose–response patterns of generally similar quantitative features with respect to amplitude and range of the stimulatory response that are either directly induced or the result of compensatory biological processes following an initial disruption in homeostasis" [101].

1.9   Conclusion

The ageing process is driven by a lifelong accumulation of molecular damage, resulting in a gradual increase in the fraction of cells carrying defects. After sufficient time has passed, the increasing levels of these defects interfere with both the performance and functional reserves of tissues and organs, resulting in a breakdown of self-organizing system and a reduced ability to adapt to the environment. It follows age-related frailty, disability, and disease. Maintenance mechanisms slow the rate of damage accumulation.

Genetics, epigenetics, sex and gender, socio-economic and educational status, chance and life circumstances, nutrition and physical activity, stress management and social support, pathogen load, all contribute to modulating positively or negatively the maintenance mechanisms. Different combinations of these factors create the possibility to avoid age-related pathologies and become centenarian. As an example, in case of nutrition, a poor diet containing excess sugar and saturated fats contributes directly to the burden of damage with which cells have to deal, whereas a Mediterranean-style diet may contribute protective factors such as dietary antioxidants, a reduced amount of animal proteins and a low glycaemic index.

There is no convincing explanation for the cause of death in apparently healthy elderly, including centenarians. The expression "dying of old age" reveals our ignorance of the biological basis of this death. Certainly, the cessation of the respiratory and circulatory functions quickly causes irreversible damage to brain, but this does not solve the problem but only moves it. The autopsy examination of a centenarian revealed stereotyped biological changes and perhaps even pathological changes that were only symptomatic or even asymptomatic in life. To date, we are not able to give a definition of "dying of old age" in biological terms. In the absence of an acute disease or a devastating chronic disease, the ageing process affects all tissues and cells, resulting in a clear progressive change for which the risk of an increased probability of dying, which defines old age, corresponds to biological and physiological variations observable empirically, without any clue about the mechanism (https://pimm.wordpress.com/2007/02/06/dying-of-old-age-an-unfounded-myth-rando-paper-box-2/). One can only state that physiological death occurs when the system represented by a living being becomes incapable of keeping itself in a stationary state in the inorganic world in which it is immersed. Since the functionality of homeostatic mechanisms ultimately depends on systems that are encoded by the genetics and modulated by all the factors quoted in the previous paragraphs, it is not surprising that different individuals have a different lifespan and some become centenarians.

References

1. Ruby JG, Wright KM, Rand KA, Kermany A, Noto K, Curtis D, et al. Estimates of the heritability of human longevity are substantially inflated due to assortative mating. Genetics. 2018;210:1109–24.
2. Avery P, Barzilai N, Benetos A, Bilianou H, Capri M, Caruso C, et al. Ageing, longevity, exceptional longevity and related genetic and non genetic markers: panel statement. Curr Vasc Pharmacol. 2014;12:659–61.
3. Dong X, Milholland B, Vijg J. Evidence for a limit to human lifespan. Nature. 2016;538:257–9.
4. Barbi E, Lagona F, Marsili M, Vaupel JW, Wachter KW. The plateau of human mortality: demography of longevity pioneers. Science. 2018;360:1459–61.
5. Gavrilov LA, Gavrilova NS. Season of birth and exceptional longevity: comparative study of American centenarians, their siblings, and spouses. J Aging Res. 2011;2011:104616.
6. Accardi G, Caruso C. Causality and chance in ageing, age-related diseases and longevity. In: Accardi G, Caruso C, editors. Updates in pathobiology: causality and chance in ageing, age-related diseases and longevity. Palermo University Press; 2017. p. 13–23.
7. Caruso C, Passarino G, Puca A, Scapagnini G. "Positive biology": the centenarian lesson. Immun Ageing. 2012;9:5.
8. Passarino G, De Rango F, Montesanto A. Human longevity: genetics or lifestyle? It takes two to tango. Immun Ageing. 2016;13:12.
9. Lutz W, Kebede E. Education and health: redrawing the Preston curve. Popul Dev Rev. 2018;44:343–61.
10. Christensen K, Doblhammer G, Rau R, Vaupel JW. Ageing populations: the challenges ahead. Lancet. 2009;374:1196–208.
11. Farrelly C. 'Positive biology' as a new paradigm for the medical sciences. Focusing on people who live long, happy, healthy lives might hold the key to improving human well-being. EMBO Rep. 2012;13:186–8.
12. Luzzatto L, Pandolfi PP. Causality and chance in the development of cancer. N Engl J Med. 2015;373:84–8.
13. Monod J. Le hasard et la nécessité: Essai sur la philosophie naturelle de la biologie moderne, Éditions du Seuil, coll. Points Essais. 1970.
14. Kirkwood TB, Feder M, Finch CE, Franceschi C, Globerson A, Klingenberg CP, et al. What accounts for the wide variation in life span of genetically identical organisms reared in a constant environment? Mech Ageing Dev. 2005;126:439–43.
15. Kirkwood TB. Understanding ageing from an evolutionary perspective. J Intern Med. 2008;263:117–27.
16. Kaern M, Elston TC, Blake WJ, Collins JJ. Stochasticity in gene expression: from theories to phenotypes. Nat Rev Genet. 2005;6:451–64.
17. Losick R, Desplan C. Stochasticity and cell fate. Science. 2008;320:65–8.
18. Beltrán-Sánchez H, Finch CE, Crimmins EM. Twentieth century surge of excess adult male mortality. Proc Natl Acad Sci U S A. 2015;112:8993–8.
19. Candore G, Balistreri CR, Colonna-Romano G, Lio D, List F, Vasto S, et al. Gender-related immune-inflammatory factors, age-related diseases, and longevity. Rejuvenation Res. 2010;13:292–7.

20. Vina J, Gambini J, Lopez-Grueso R, Abdelaziz KM, Jove M, Borras C. Females live longer than males: role of oxidative stress. Curr Pharm Des. 2011;17:3959–65.
21. Caruso C, Accardi G, Virruso C, Candore G. Sex, gender and immunosenescence: a key to understand the different lifespan between men and women? Immun Ageing. 2013;10:20.
22. Caruso C, Vasto S. Immunity and aging. In: Ratcliffe MJH, editor. Encyclopedia of immunobiology, vol. 5. Oxford: Academic; 2016. p. 127–32.
23. Accardi G, Caruso C. Immune-inflammatory responses in the elderly: an update. Immun Ageing. 2018;15:11.
24. Finch CE, Crimmins EM. Inflammatory exposure and historical changes in human life-spans. Science. 2004;305:1736–9.
25. Balistreri CR, Candore G, Caruso C. Role of TLR polymorphisms in aging and age-related diseases. In: Fulop T, Franceschi C, Hirokawa K, Pawelec G, editors. Handbook of immunosenescence. Cham: Springer; 2019. In press.
26. Licastro F, Candore G, Lio D, Porcellini E, Colonna-Romano G, Franceschi C, et al. Innate immunity and inflammation in ageing: a key for understanding age-related diseases. Immun Ageing. 2005;2:8.
27. Candore G, Caruso C, Jirillo E, Magrone T, Vasto S. Low grade inflammation as a common pathogenetic denominator in age-related diseases: novel drug targets for anti-ageing strategies and successful ageing achievement. Curr Pharm Des. 2010;16:584–96.
28. Arai Y, Martin-Ruiz CM, Takayama M, Abe Y, Takebayashi T, Koyasu S, et al. Inflammation, but not telomere length, predicts successful ageing at extreme old age: a longitudinal study of semi-supercentenarians. EBioMedicine. 2015;2:1549–58.
29. Storci G, De Carolis S, Papi A, Bacalini MG, Gensous N, Marasco E, et al. Genomic stability, anti-inflammatory phenotype, and up-regulation of the RNAseH2 in cells from centenarians. Cell Death Differ. 2019. https://doi.org/10.1038/s41418-018-0255-8.
30. De la Fuente M, Miquel J. An update of the oxidation-inflammation theory of aging: the involvement of the immune system in oxi-inflammaging. Curr Pharm Des. 2009;15:3003–26.
31. Biswas SK. Does the Interdependence between oxidative stress and inflammation explain the antioxidant paradox? Oxidative Med Cell Longev. 2016;2016:5698931.
32. Barja G. Rate of generation of oxidative stress-related damage and animal longevity. Free Radic Biol Med. 2002;33:1167–72.
33. Sohal RS, Mockett RJ, Orr WC. Mechanisms of aging: an appraisal of the oxidative stress hypothesis. Free Radic Biol Med. 2002;33:575–86.
34. Liguori I, Russo G, Curcio F, Bulli G, Aran L, Della-Morte D, et al. Oxidative stress, aging, and diseases. Clin Interv Aging. 2018;13:757–72.
35. Bernadotte A, Mikhelson VM, Spivak IM. Markers of cellular senescence. Telomere shortening as a marker of cellular senescence. Aging (Albany NY). 2016;8:3–11.
36. Rizvi SI, Maurya PK. Alterations in antioxidant enzymes during aging in humans. Mol Biotechnol. 2007;37:58–61.
37. Gill R, Tsung A, Billiar T. Linking oxidative stress to inflammation: toll-like receptors. Free Radic Biol Med. 2010;48:1121–32.
38. Marinin F. Signaling by ROS drives inflammasome activation. Eur J Immunol. 2010;40:595–603.
39. Sharma A, Tate M, Mathew G, Vince JE, Ritchie RH, de Haan JB. Oxidative stress and NLRP3-inflammasome activity as significant drivers of diabetic cardiovascular complications: therapeutic implications. Front Physiol. 2018;9:114.
40. Halliwell B. Antioxidant defence mechanisms: from the beginning to the end (of beginning). Free Radic Res. 1999;31:261–72.
41. Paolisso G, Barbieri M, Bonafè M, Franceschi C. Metabolic age modelling: the lesson from centenarians. Eur J Clin Invest. 2000;30:888–94.
42. Mecocci P, Polidori MC, Troiano L, Cherubini A, Cecchetti R, Pini G, et al. Plasma antioxidants and longevity: a study on healthy centenarians. Free Radic Biol Med. 2000;28:1243–8. Erratum in: Free Radic Biol Med. 2000;29:486.

43. Polidori MC, Mariani E, Baggio G, Deiana L, Carru C, Pes GM, et al. Different antioxidant profiles in Italian centenarians: the Sardinian peculiarity. Eur J Clin Nutr. 2007;61:922–4.
44. Cevenini E, Caruso C, Candore G, Capri M, Nuzzo D, Duro G, et al. Age-related inflammation: the contribution of different organs, tissues and systems. How to face it for therapeutic approaches. Curr Pharm Des. 2010;16:609–18.
45. Vasto S, Buscemi S, Barera A, Di Carlo M, Accardi G, Caruso C. Mediterranean diet and healthy ageing: a Sicilian perspective. Gerontology. 2014;60:508–18.
46. Aiello A, Accardi G, Candore G, Gambino CM, Mirisola M, Taormina G, et al. Nutrient sensing pathways as therapeutic targets for healthy ageing. Expert Opin Ther Targets. 2017;21:371–80.
47. Leonardi GC, Accardi G, Monastero R, Nicoletti F, Libra M. Ageing: from inflammation to cancer. Immun Ageing. 2018;15:1.
48. Dato S, Crocco P, D'Aquila P, de Rango F, Bellizzi D, Rose G, et al. Exploring the role of genetic variability and lifestyle in oxidative stress response for healthy aging and longevity. Int J Mol Sci. 2013;14:16443–72.
49. Longo VD, Antebi A, Bartke A, Barzilai N, Brown-Borg HM, Caruso C, et al. Interventions to slow aging in humans: are we ready? Aging Cell. 2015;14:497–510.
50. Partridge L. The new biology of ageing. Philos Trans R Soc Lond Ser B Biol Sci. 2010;365:147–54.
51. Fontana L, Kennedy BK, Longo VD, Seals D, Melov S. Medical research: treat ageing. Nature. 2014;511:405–7.
52. Lawrence T. The nuclear factor NF-kappaB pathway in inflammation. Cold Spring Harb Perspect Biol. 2009;1:a001651.
53. Soultoukis GA, Partridge L. Dietary protein, metabolism, and aging. Annu Rev Biochem. 2016;85:5–34.
54. Blumenthal HT. The aging-disease dichotomy: true or false? J Gerontol A Biol Sci Med Sci. 2003;58:138–45.
55. Schumacher B, van der Pluijm I, Moorhouse MJ, Kostea T, Robinson AR, Suh Y, et al. Delayed and accelerated aging share common longevity assurance mechanisms. PLoS Genet. 2008;4:e1000161.
56. Garinis GA, van der Horst GT, Vijg J, Hoeijmakers JH. DNA damage and ageing: new-age ideas for an age-old problem. Nat Cell Biol. 2008;10:1241–7.
57. Johnson SC, Rabinovitch PS, Kaeberlein M. mTOR is a key modulator of ageing and age-related disease. Nature. 2013;493:338–45.
58. Di Bona D, Accardi G, Virruso C, Candore G. Caruso l. Association between genetic variations in the insulin/insulin-like growth factor (Igf-1) signaling pathway and longevity: a systematic review and meta-analysis. Curr Vasc Pharmacol. 2014;12:674–81.
59. Siddle K. Signalling by insulin and IGF receptors: supporting acts and new players. J Mol Endocrinol. 2011;47:R1–10.
60. Houtkooper RH, Pirinen E, Auwerx J. Sirtuins as regulators of metabolism and healthspan. Nat Rev Mol Cell Biol. 2012;13:225–38.
61. Wąsroba M, Szukiewicz D. The role of sirtuins in aging and age related diseases. Adv Med Sci. 2016;61:52–62.
62. Martin DE, Hall MN. The expanding TOR signaling network. Curr Opin Cell Biol. 2005;17:158–66.
63. Galluzzi L, Baehrecke EH, Ballabio A, Boya P, Bravo-San Pedro JM, Cecconi F, et al. Molecular definitions of autophagy and related processes. EMBO J. 2017;36:1811–36.
64. Puca AA, Spinelli C, Accardi G, Villa F, Caruso C. Centenarians as a model to discover genetic and epigenetic signatures of healthy ageing. Mech Ageing Dev. 2018;174:95–102.
65. Accardi G, Aprile S, Candore G, Caruso C, Cusimano R, Cristaldi L, et al. Genotypic and phenotypic aspects of longevity: results from a Sicilian survey and implication for the prevention and the treatment of age-related diseases. Curr Pharm Des. 2019;25:228–35.
66. Aiello A, Accardi G, Candore G, Carruba G, Davinelli S, Passarino G, et al. Nutrigerontology: a key for achieving successful ageing and longevity. Immun Ageing. 2016;13:17.

67. Aiello A, Caruso C, Accardi G. Slow-aging diets. In: Danan G, Matthew Dupre E, editors. Encyclopedia of gerontology and population aging. Springer; 2019. In press. https://doi.org/10.1007/978-3-319-69892-2_134-1.
68. Mirzaei H, Suarez JA, Longo VD. Protein and amino acid restriction, aging and disease: from yeast to humans. Trends Endocrinol Metab. 2014;25:55866.
69. Fontana L, Partridge L, Longo VD. Extending healthy life span—from yeast to humans. Science. 2010;328:321–6.
70. Most J, Tosti V, Redman LM, Fontana L. Calorie restriction in humans: an update. Ageing Res Rev. 2017;39:36–45.
71. Speakman JR, Mitchell SE. Caloric restriction. Mol Asp Med. 2011;32:159–221.
72. Redman LM, Kraus WE, Bhapkar M, Das SK, Racette SB, Martin CK, et al. Energy requirements in nonobese men and women: results from CALERIE. Am J Clin Nutr. 2014;99:71–8.
73. Rakowski W, Mor V. The association of physical activity with mortality among older adults in the Longitudinal Study of Aging (1984-1988). J Gerontol. 1992;47:M122–9.
74. Warburton DER, Charlesworth S, Ivey A, Nettlefold L, Bredin SSD. A systematic review of the evidence for Canada's Physical Activity Guidelines for Adults. Int J Behav Nutr Phys Act. 2010;7:39.
75. Samitz G, Egger M, Zwahlen M. Domains of physical activity and all-cause mortality: systematic review and dose-response meta-analysis of cohort studies. Int J Epidemiol. 2011;40:1382–400.
76. Reimers CD, Knapp G, Reimers AK. Does physical activity increase life expectancy? A review of the literature. J Aging Res. 2012;2012:243958.
77. López-Otín C, Blasco MA, Partridge L, Serrano M, Kroemer G. The hallmarks of aging. Cell. 2013;153:1194–217.
78. Garatachea N, Pareja-Galeano H, Sanchis-Gomar F, Santos-Lozano A, Fiuza-Luces C, Morán M, et al. Exercise attenuates the major hallmarks of aging. Rejuvenation Res. 2015;18:57–89.
79. Rebelo-Marques A, De Sousa Lages A, Andrade R, Ribeiro CF, Mota-Pinto A, Carrilho F, et al. Aging hallmarks: the benefits of physical exercise. Front Endocrinol (Lausanne). 2018;9:258.
80. Radák Z, Naito H, Kaneko T, Tahara S, Nakamoto H, Takahashi R, et al. Exercise training decreases DNA damage and increases DNA repair and resistance against oxidative stress of proteins in aged rat skeletal muscle. Pflügers Arch. 2002;445:273–8.
81. Gomez-Cabrera M-C, Domenech E, Viña J. Moderate exercise is an antioxidant: upregulation of antioxidant genes by training. Free Radic Biol Med. 2008;44:126–31.
82. Leick L, Lyngby SS, Wojtasewski JF, Pilegaard H. PGC-1α is required for training-induced prevention of age-associated decline in mitochondrial enzymes in mouse skeletal muscle. Exp Gerontol. 2010;45:336–42.
83. Puterman E, Lin J, Blackburn E, O'Donovan A, Adler N, Epel E. The power of exercise: buffering the effect of chronic stress on telomere length. PLoS One. 2010;5:e10837.
84. Grazioli E, Dimauro I, Mercatelli N, Wang G, Pitsiladis Y, Di Luigi L, et al. Physical activity in the prevention of human diseases: role of epigenetic modifications. BMC Genomics. 2017;18:802.
85. Perry CG, Lally J, Holloway GP, Heigenhauser GJ, Bonen A, Spriet LL. Repeated transient mRNA bursts precede increases in transcriptional and mitochondrial proteins during training in human skeletal muscle. J Physiol. 2010;588:4795–810.
86. Nakajima K, Takeoka M, Mori M, Hashimoto S, Sakurai A, Nose H, et al. Exercise effects on methylation of ASC gene. Int J Sports Med. 2010;31:671–5.
87. Kim YA, Kim YS, Oh SL, Kim H-J, Song W. Autophagic response to exercise training in skeletal muscle with age. J Physiol Biochem. 2013;69:697–705.
88. Yarasheski KE, Pak-Loduca J, Hasten DL, Obert KA, Brown MB, Sinacore DR. Resistance exercise training increases mixed muscle protein synthesis rate in frail women and men $\geq$ 76 yr old. Am J Physiol Endocrinol Metab. 1999;277:E118–25.
89. Woods JA, Wilund KR, Martin SA, Kistler BM. Exercise, inflammation and aging. Aging Dis. 2012;3:130–40.

90. Kokkinos P, Sheriff H, Kheirbek R. Physical inactivity and mortality risk. Cardiol Res Pract. 2011;2011:924945.
91. Wade KH, Richmond RC, Davey Smith G. Physical activity and longevity: how to move closer to causal inference. Br J Sports Med. 2018;52:890– .
92. Kujala UM. Is physical activity a cause of longevity? It s not as straightforward as some would believe. A critical analysis. Br J Sports Med. 2018;52:914–8.
93. Brefczynski-Lewis JA, Lutz A, Schaefer HS, Levinson DB, Davidson RJ. Neural correlates of attentional expertise in long-term meditation practitioners. Proc Natl Acad Sci U S A. 2007;104:11483–8.
94. Gard T, Hölzel BK, Lazar SW. The potential effects of meditation on age-related cognitive decline: a systematic review. Ann N Y Acad Sci. 2014;1307:89–103.
95. Sharma H. Meditation: process and effects. Ayu. 2015;36:233–7.
96. Rosenkranz MA, Lutz A, Perlman DM, Bachhuber DR, Schuyler BS, MacCoon DG, et al. Reduced stress and inflammatory responsiveness in experienced meditators compared to a matched healthy control group. Psychoneuroendocrinology. 2016;68:117–25.
97. Epel E, Blackburn E, Lin J, Dhabhar F, Adler N, Morrow JD, et al. Accelerated telomere shortening in response to exposure to life stress. Proc Natl Acad Sci U S A. 2004;101:17312–5.
98. Bakaysa SL, Mucci LA, Slagboom E, Boomsma DI, McClearn GE, Johansson B, et al. Telomere length predicts survival independent of genetic influences. Aging Cell. 2007;6:769–74.
99. Damjanovic AK, Yang Y, Glaser R, Kiecolt-Glaser JK, Nguyen H, Laskowski B, et al. Accelerated telomere erosion is associated with a declining immune function of caregivers of Alzheimer's disease patients. J Immunol. 2007;179:4249–54.
100. Hathcock KS, Chiang Y, Hodes RJ. In vivo regulation of telomerase activity and telomere length. Immunol Rev. 2005;205:104–13.
101. Calabrese J, Baldwin LA. Defining hormesis. Hum Exp Toxicol. 2002;21:91–7.

Phenotypic Aspects of Longevity

2

Giulia Accardi, Mattia Emanuela Ligotti,
and Giuseppina Candore

2.1 Introduction

The phenotype is the set of morphofunctional features of an organism. These are the result of the interaction between genotype and environment. Classically, longevity phenotype is considered to be the outcome of a positive combination between genetic (25%), epigenetic (25%) and lifestyle factors (50%) that lead an organism to reach advanced age. Indeed, the contribution of genetics in human longevity is estimated around 25%, but this percentage increases with age. In fact, genetics is considered negligible up to the age of 55–60 [1] and varies depending on the analysed cohort, from 15% in historic Alpine communities [2] to 33% in Swedish twins [3].

The exceptional longevity is a trait running in the family. In addition, its heritability percentages increase with age and depend on sex (up to 33% for women, 48% for men) [1, 4].

Thus, ageing is a multifactorial process that results in survival of individuals over the average life span characteristic of their population. Concerning the phenotype, it is important to distinguish between chronological age (CA) and biological age (BA).

The CA is the age of an individual expressed as time that has elapsed since birth. Instead, the BA reflects the residual functional capacity of an organism and often differs from CA.

This means that faster or slower the ageing process, greater is the distance between CA and BA.

G. Accardi (✉)· M. E. Ligotti · G. Candore
Laboratory of Immunopathology and Immunosenescence, Department of Biomedicine,
Neurosciences and Advanced Diagnostics, University of Palermo, Palermo, Italy
e-mail: giulia.accardi@unipa.it; mattiaemanuela.ligotti@unipa.it; giuseppina.candore@unipa.it

© Springer Nature Switzerland AG 2019
C. Caruso (ed.), *Centenarians*, https://doi.org/10.1007/978-3-030-20762-5_2

Table 2.1 Criteria for ageing biomarker by the American Federation for Aging Research (2011)

1	It must predict the rate of ageing. In other words, it would tell exactly where a person is in their total life span. It must be a better predictor of life span than chronological age
2	It must monitor a basic process that underlies the ageing process, not the effects of disease
3	It must be able to be tested repeatedly without harming the person. For example, a blood test or an imaging technique
4	It must be something that works in humans and in laboratory animals, such as mice. This is so that it can be tested in laboratory animals before being validated in humans

Therefore, one approach to determine if a person is older or younger than expected is to make a comparison with general population, using validated parameters with known range of values, the so-called biomarkers. These are "biological parameters of an organism that either alone or in some multivariate composite will, in the absence of disease, better predict functional capability at some late age" (see Table 2.1).

So, for the growing number of old people worldwide, the importance to identify easy and low-cost biomarkers of ageing is crucial.

Based on the improvements in sanitation, in medical care and in diet behaviours, centenarian populations have grown rapidly during the last years. According to Bonarini [5], in Italy, there were 0.062/10,000 centenarians in 1901 whereas in 2009 they were 2.4/10,000. Considering that their younger contemporary subjects are exposed to the same environmental conditions, the main survival advantage may be probably due to the favourable interaction between genetic background and environment, rather than single factors [6].

Worldwide, there are four geographical areas with high rates of centenarians, the so-called blue zone (BZ): Ogliastra in Sardinia, Okinawa in Japan, the Nicoya peninsula in Costa Rica and the island of Ikaria in Greece. These are defined as a "rather limited and homogenous geographical area where the population shares the same lifestyle and environment and its longevity has been proved to be exceptionally high" [7]. In addition, a fifth zone has been suggested, the Loma Linda town where live the Seventh-day Adventists, by National Geographic (https://www.nationalgeographic.com/travel/happiest-places/blue-zones-california-photos/).

Supercentenarians, i.e. people aged 110 years or above, represent a more selected subgroup compared with centenarians, but they are extremely rare (37 validated living supercentenarians worldwide, among which 6 Italians, to date) [8], so difficult to analyse.

A noteworthy study of 32 US supercentenarians (although no age-validated) showed that about 41% needed minimal or no assistance. Moreover, the prevalence of cardiovascular diseases (CVDs) and stroke was low among them [9]. This datum was confirmed in a study on 12 age-validated Japanese supercentenarians, where it was found that 83% of them did not report the most common age-related morbidities (e.g., CVD, stroke, cancer) up to age 105 [10].

Regarding neurodegenerative diseases, autopsy of four supercentenarians showed well-preserved brain with mild neuropathological alterations. Furthermore, from the analysis of apolipoprotein E (APOE) gene, no ε4 allele was detected [11], corroborating the hypothesis that long-living individuals (LLIs), i.e. people belonging to the five percentile of the survival curve, may have some protective genetic factors that give them a survival advantage (see Chap. 6).

In this chapter, we will overlook the main phenotypic aspects of longevity, focusing on recent scientific data about LLIs, including centenarians. The aim is to realize, through a positive biology approach, i.e. understanding the causes of centenarian positive phenotype, how to prevent and/or postpone age-related diseases.

2.2   Haematochemical Values

Haematochemical biomarkers are the easiest and cheapest tools for diagnosis and prognosis of many diseases. Periodically, the World Health Organization updates the reference range, depending on new studies and findings to classify people as healthy or unhealthy. Scientific reports show that some values vary during ageing, whereas others do not seem to significantly change [12].

Although, in some cases different range exists depending on age and gender, there are no specific parameters for oldest old.

However, scientific reports demonstrate that although the majority of values are similar comparing adult and centenarians, some differences exist (e.g., urea nitrogen and creatinine increase in centenarians, whereas platelets count, serum protein and vitamin B12 and D decrease) [13]. For sure, it is necessary to consider the presence of inflamm-ageing, the chronic, low-grade, progressive inflammatory status affecting old people (see below) that influences haematochemical parameters, including erythrocyte sedimentation rate and C-reactive protein (CRP). About lipids (triglyceride, total cholesterol, low-density lipoprotein and high-density lipoprotein or HDL) exist contrasting results [14]. In the recent survey on Sicilian LLIs (mean age 101.3 ± 4.9), preliminary unpublished results suggest no differences concerning lipid profile, glucose and insulin levels when compared with young (mean age 30.7 ± 4.8) individuals, whereas creatinine was increased (Ciaccio, Caldarella and Caruso, unpublished observations).

2.3   Age-Related Diseases in Centenarians

Centenarians, despite their so advanced age, show relatively good health, postponing or escaping the major age-related diseases, including CVDs, dementia and cancer.

Based on their morbidity profiles, they can be divided into three groups: survivors (onset of age-related disease before 80th year), delayers (onset of age-related disease after 80th year) and escapers (without common age-related disease before 100th year) [15]. Although the percentages of these groups may vary among different cohorts, a common datum is that when compared with their younger contemporary subjects, the centenarians seem to show less severe diseases [16].

More recently, centenarians have been divided into low and high performers, depending on the ability to avoid or postpone major age-related diseases. The high-performing centenarians are those free from diseases or with a very late onset of diseases. The low-performing (the majority) suffer for these pathologies so they need care from relatives or nurse facilities. One possible explanation of this difference is linked to the different expression of some hallmarks of ageing, in particular the telomere length (greater in high-performing) (see Chap. 8) and the level of immunosenescence (more pronounced in low-performing) (see Chap. 3) [17]. Considering the fast increase of old people worldwide, including the elderly with chronic diseases and/or with sensory and cognitive changes, it is not surprising that the extraordinary resilience of centenarians has attracted the interest of researchers in the last decades.

Literature data from different populations show a broad variation about the prevalence of chronic diseases in centenarians, probably due to differences in gender, ethnicity, sampling methods and to the lack of an age-matched control group. However, there is consensus that centenarians constitute an informative population for studying genetic and environmental variables, and their interactions, involved in longevity phenotype.

Numerous studies show a close association between advanced age and the pathogenesis of cancer [18–20] although the major risk of cancer in the elderly is represented by long-term exposure to relevant carcinogenic factors that lead to an increase in the cancer rate with age. Indeed, despite the advances in early detection and in treatment, this pathology is one of the most common causes of death in individuals older than 65 [21], representing an ever more important public health problem. For example, according to the United States Cancer Statistics [22], in 2015, the age-specific rate of new cancers per 100,000 people increased from 1574 in 65–69 years old group to 2200 in 80–84 and decreased to 2020 in over 85.

Several studies have focused attention on the relationship between longevity and cancer, analysing nonagenarians and centenarians. In particular, autopsy records revealed that centenarians are characterized by a significant delayed age of diagnosis of some cancers [23] and a lowest rate of metastasis [24] when compared with the elderly. These data were confirmed by an American longitudinal cohort study that explored the prevalence for specific types of cancer in nonagenarians and centenarians. From the analysis of 1143 subjects (90–119 years) compared with surveillance, epidemiology and end results (SEER) data from the National Cancer Institute, collecting information from the entire US population, it emerged that 20.3% (20% female, 22% male) centenarians had a history of non-skin cancer, compared with 34.4% SEER probability of developing non-skin cancer. The most common cancers in centenarians were prostate (11.7%), breast (8.2%) and colon (5.7%), with the mean age at diagnosis of 80.5 years compared with 63.2 years in SEER data [25]. Although the prevalence varies with cancer type, some are very rare among centenarians, suggesting that these individuals may have a protective genetic advantage to resist or delay cancer development.

One explanation may be related to the different gene dosage of Arf/p53. The higher the number of copies of these genes (so higher is the gene product), the greater the pro-longevity and anti-ageing effect. They favour longevity and slow ageing through protection from DNA damage that leads to a tumour suppressor activity and reacting to stressors. Moreover, their increased expression has a role in tissue homeostasis and regeneration reducing the exhaustion of the pool of stem cells [26].

2.4   Oxidative Stress and Inflammation

Changes associated with ageing also affect the immune-inflammatory responses as shown by decline in immune function and increase in the systemic pro-inflammatory status, i.e. immunosenescence [27]. Immunosenescence is linked not only to the functional decline associated with the passage of time but also to antigen burden to which an individual has been exposed during lifetime. Therefore, a senescent immune system is characterized by continuous reshaping and shrinkage of the immune repertoire by persistent antigenic challenges. These changes lead to a poor response to newly encountered microbial antigens, including vaccines, as well as to a shift of the immune system towards an inflammatory Th2 profile. It is well known that the pro-inflammatory status, called inflamm-ageing, is characterized by elevated levels of pro-inflammatory cytokines, such as interleukin (IL)-6, IL-1ß and tumor necrosis factor (TNF)-α [27]. This leads to an imbalance between pro- and anti-inflammatory status, contributing to the increase in susceptibility of the elderly to new infections. However, the mechanisms underlying are not completely understood (for immunophenotype see Chap. 3).

In this context, centenarians show an increase in many inflammatory molecules comparing to adults, either pro-inflammatory such as IL-6, IL-18 and IL-15, CRP, fibrinogen, serum-amyloid A, Von Willebrand factor, resistin and leukotrienes [14], or anti-inflammatory such as adiponectin, transforming growth factor-β1, IL-1 receptor antagonist, cortisol and arachidonic acid derivatives, such as hydroxyeicosatetraenoic acids and epoxyeicosatrienoicacids [28–30]. The increase in pro-inflammatory cytokines can provide protection to infectious diseases, but also can lead to an inflammatory status, correlated to inflamm-ageing. On the other hand, this condition is compensated by a concomitant activation of anti-inflammatory responses as a whole working as anti-inflammageing components [31]. This suggests that inflamm-ageing is likely an adaptive mechanism that occurs throughout life [32] and may coexist with longevity especially if counterbalanced by an anti-inflammatory component.

Other authors, instead, support the oxi-inflammageing theory, according to which the oxidative stress is one of the major determinants of low-grade inflammatory status affecting the elderly [33].

As stated in this theory, mitochondria produce elevated levels of reactive oxygen species (ROS), highly reactive molecules that may induce oxidative damage to proteins, lipids and nucleic acids in response to different environmental stimuli, including inflammation [34]. The increase of ROS is correlated to damage-associated molecular pattern (DAMPs) release, endogenous nuclear or cytosolic molecules released from injured and dying cells [35]. DAMPs are able to activate inflammasome, as NLRP-3, through specific receptors, promoting the production of pro-inflammatory cytokines, such as IL-1ß, IL-6 and TNF-α. In turn, these cytokines activate innate immune cells, further potentiating ROS production, in a vicious cycle [36].

To maintain ROS levels at low and counteract oxidative stress, cells use various antioxidants, including enzymes as superoxide dismutase, catalase and glutathione peroxidase, and non-enzymatic molecules as uric acid, ascorbic acid, glutathione, Coenzyme Q10 (CoQ10), and nutritional factors, including vitamin C and E. However, with ageing oxidative damage accumulation and decrease in the antioxidant ability of the cell occur, resulting in a status of altered homeostasis that exacerbates the ageing-associated inflammatory status, through nuclear factor-κB activation [36, 37]. This condition of vulnerability exposes the elderly to a high risk of disease and ultimately to death [38].

Within this scenario, data from the literature report both increase and decrease of oxidative stress in LLIs. In particular, in most studies it was seen that centenarians had lower levels of lipid peroxides, markers of oxidative stress, and higher plasma levels of vitamin E than elderly controls [39–42], suggesting that they may be genetically better equipped to contrast oxidative stress.

However, a recent Japanese study demonstrated that centenarians show a decrease in vitamin C and an increase in the oxidized form of CoQ10 when compared with the elderly [43], indicating an increased oxidative stress in these individuals.

In order to clarify the role of the oxidative stress in human life-span, Conte and co-workers [44] analysed the potential link between plasma levels of mitokines (molecules produced in response to mitochondrial stress that mediate a survival signal [45]), e.g., fibroblast growth factor 21 and growth differentiation factor 15, and functional and biochemical parameters in a cohort of Italian subjects aged from 21 to 113 years. It was seen that circulating levels of mitokines increased with age and they were inversely related to functional and biochemical parameters as handgrip strength, HDL cholesterol, triglycerides, body mass index (BMI) and uric acid in elderly, and with survival in LLIs.

Furthermore, emerging evidence from studies in animal models has linked a mild mitochondrial dysfunction with an increase of life-span, questioning the role of oxidative stress in longevity and reinforcing the hypothesis of mitohormesis, a response of mitochondria to stresses that activate cytosolic signalling pathways, modulating gene expression and leaving the cell less vulnerable to subsequent perturbations [46].

These controversial data suggest that ageing is associated with a progressive mitochondrial dysfunction due to oxidative stress, and leading to it in a vicious cycle, which must be considered in longevity studies.

In this context, nutrition is an important modulator of stress levels. Hypernutrition can augment cell damages caused by oxidative stress, triggering inflammatory intracellular signalling pathways. At the same time, a diet rich in nutraceuticals, as the Mediterranean one, is able to stimulate a variety of anti-inflammatory and antioxidant molecular mechanisms, contributing to decrease systemic inflammation and oxidative status (see Chaps. 5 and 10). Therefore, the difference in dietary pattern, both geographical and historical, could partly explain discrepancies among centenarians features worldwide.

2.5   Psychological and Cognitive Function

The deterioration of cognitive functions, including memory, comprehension and learning capacity, is one of the major consequences of ageing process. Today, 50 million people live with dementia, and this number is estimated to exponentially increase over the next years, predominantly in developing countries [47].

The maintenance of good cognitive function is essential for the quality of life not only for the elderly but also for their families that become spectators of the personality and behavioural changes of their relatives. Therefore, it is needed to identify protective genetic and environmental factors towards cognitive decline.

Several studies analysed the cognitive status in centenarians and oldest old, but the rate of impairment reported in the literature varies widely, ranging from 7% [48] to 77.5% [49] of the analysed populations. There are many reasons for this variability, some related to sampling method (sample sizes, inclusion and/or exclusion criteria), others related to testing procedures or to individual characteristics (gender, ethnicity, education and lifestyle).

The most used test for the assessment of cognitive function is the mini-mental state examination (MMSE), which often may not be suitable for oldest old, as it requires the ability to write, read and hear [50]. Nevertheless, a percentage of centenarians, albeit low in some studies, are able to get a good MMSE score, suggesting that the preservation of high cognitive functioning in late life is a complex phenomenon influenced by gene–environment interactions. In this context, the APOEε4 gene variant has been related to more severe cognitive impairment and to development of Alzheimer's disease [51]. Some studies have proposed education as a possible factor able to counteract the negative effect of APOEε4 allele on cognitive decline. For example, Ishioka et al. [52] have found that there was no difference between the ε4 carriers and non-carriers among women centenarians with higher educational levels.

Another factor that seems to lead to a cognitive decline is the carotid atherosclerosis. Kawasaki et al. [53] have seen that old people, including centenarians, with carotid artery plaque had lower MMSE scores compared with those without. Based on these data, the authors proposed the prevention of atherosclerosis as a strategy that can contribute to the achievement of successful brain ageing.

As well as the evaluation of physical and cognitive functions, the analysis of the psychological state, and in particular how they adapt to the physiological age-related changes, is important. In a recent study, the well-being of centenarians was compared in two very different cultural contexts, Japan and the United States, through the use of Philadelphia Geriatric Center Morale Scale, which allows to measure psychological comfort, satisfaction with social relations and attitude towards own ageing. From this cross-cultural comparison, it emerged that American centenarians had higher levels of satisfaction about social relations and psychological comfort but no difference in attitude towards own ageing [54], interpreted as cognitive well-being. Regression analyses revealed that only in Japanese centenarians, attitude towards own ageing was strongly influenced by social resources (living alone or cohabit). These results can be explained by the different cultural traditions of the two countries, which inevitably affect the perception of oneself and interpersonal relationships.

2.6   Body Composition

The somatotypes are "particular categories of body build, determined on the basis of certain physical characteristics. The three basic body types are ectomorph (thin physique), endomorph (rounded physique) and mesomorph (athletic physique)" [55]. The classification in one of these types can be made by the body composition analysis. It is the measurement of the different compartments of the human body (fat-free mass, including total or compartmental body water, bone mass, muscle mass and fat mass). Although at first sight it could seem to be strictly related to the physical condition, in terms of fitness and wellness, in reality, it is a useful tool for the evaluation of general health condition, organ function, drug treatment or diagnosis of diseases [56]. Nonetheless, it represents the starting point to develop specific personalized dietary program or age- and sex-matched protocols for population.

The limit of this in studies on centenarians is the absence of age-matched reference range. In fact, for example, if for adults BMI over 24.9 Kg/m^2 represents an index of overweight, it has been speculated a possible inverse or null relationship between overweight and obesity with mortality in old people [57, 58]. The reason seems to be associated with the protection by fat accumulation from death called obesity paradox that keeps in light the low specificity of BMI for the classification of population based on height and weight measures alone [59, 60]. This is particularly true for the elderly that often are characterized by reduction of fat-free mass (excluding total body water (TBW) often increased due to extracellular water accumulation for oedema) and redistribution of adipose tissue but with same BMI [61].

A complete analysis of body composition consists of direct and indirect measures: waist circumference, waist-to-hip ratio and waist-to-height ratio, percentage of fat-free mass, fat mass and phase angle (Pa).

The Pa is a vectorial value calculated by medically validated algorithms, resulting from two measures: resistance (Rz) and reactance (Xc) of human tissues, strictly related to body composition. The Rz is inversely proportional to TBW (the higher the Rz, the lower is TBW). The Xc is in direct proportion with cell density in tissues. So, the Pa could be considered a measure of the quality and integrity of a cell and its membrane. Lower is the Pa, higher is the cell damage, in terms of breakdown of selective membrane, hence permeability of cell (it was seen that people with low Pa have higher Na+/K+ ratio). This consequently means that the higher the Pa value, the better the healthy condition [62–64]. Thus, the identification of normal range of Pa at different age could be a measurable value of health or diseases, although its biological meaning is still not fully clear. At the moment, scientific evidences demonstrated a role in the prognosis of sarcopenia and cachexia [65]. But the reference to the normal range of values (mean 6.5°) for the oldest old does not permit a good classification of LLI, having a body composition totally different from adult or young people.

An on-going study of Sicilian LLIs (mean age of 101.5) is trying to understand the differences in body composition between young adult and LLIs to identify possible new range of values of anthropometric measurements for oldest old. Preliminary unpublished results (Accardi, Aiello and Caruso) reported a mean BMI for healthy LLIs of 24.35 Kg/m². The body composition analysis keeps in light a general hyperhydration, probably due to widespread oedema possibly linked to inflamm-ageing or kidney or heart dysfunction. Moreover, the mean value of Pa in these LLIs (3.2°) classifies them as cachectic (Pa 3.5° or lower).

2.7 Conclusion

The study of centenarians is an opportunity to understand their "secrets" for the achievement of their extreme longevity. The observation of their lifestyle and the analysis of their background are really interesting but, in reality, not so useful to delineate a unique signature of longevity. This is due to the different interactions between genetics and lifestyle and to the fact that every people is a single and not replicable individual.

Longevity is the result of genetic and epigenetic make up in combination with environmental condition, causality and chance, all the life long, from the life in utero. For this reason, it is not possible to make a standardization of the "perfect match" to become centenarian. In addition, centenarians are rare, and often, they live in big cities so in totally different conditions.

To overcome this problem, it might be more useful to analyse zone with high number of LLIs or centenarians, as the BZ where it should be possible to identify some common elements to make understandable association with longevity.

Despite these limitations, the observation of longevity phenotype could be an opportunity to identify a good strategy to achieve successful ageing and reach healthy life span. It is evident that this kind of studies requires a multidisciplinary approach, which must also be extended to the psychological aspects of ageing, in terms of subjective well-being, i.e. emotional and cognitive valuation of one's life.

References

1. Brooks-Wilson AR. Genetics of healthy aging and longevity. Hum Genet. 2013;132(12):1323–38.
2. Gögele M, Pattaro C, Fuchsberger C, Minelli C, Pramstaller PP, Wjst M. Heritability analysis of life span in a semi-isolated population followed across four centuries reveals the presence of pleiotropy between life span and reproduction. J Gerontol A Biol Sci Med Sci. 2011;66(1):26–37.
3. Ljungquist B, Berg S, Lanke J, McClearn GE, Pedersen NL. The effect of genetic factors for longevity: a comparison of identical and fraternal twins in the Swedish Twin Registry. J Gerontol A Biol Sci Med Sci. 1998;53(6):M441–6.
4. Cao H, Gerhold K, Mayers JR, Wiest MM, Watkins SM, Hotamisligil GS. Identification of a lipokine, a lipid hormone linking adipose tissue to systemic metabolism. Cell. 2008;134: 933–44.
5. Bonarini F. Il numero dei centenari in Italia. Working Paper Series; 2009 Feb 20. N. 4 (in Italian).
6. Villa F, Spinelli CC, Puca AA. Diet and longevity phenotype. In: Molecular basis of nutrition and aging: a volume in the molecular nutrition series. Elsevier Inc.; 2016. p. 31–9.
7. Poulain M, Pes GM. The Blue Zones: areas of exceptional longevity around the world. In: Vienna yearbook of population research. Vol. 11. 2013. p. 87–108.
8. Gerontology Research Group [Internet]. Numbers of validated living supercentenarians. http://www.grg.org/SC/WorldSCRankingsList.html. Accessed Dec 2018.
9. Schoenhofen EA, Wyszynski DF, Andersen S, Pennington J, Young R, Terry DF, et al. Characteristics of 32 supercentenarians. J Am Geriatr Soc. 2006;54(8):1237–40.
10. Willcox DC, Willcox BJ, Wang NC, He Q, Rosenbaum M, Suzuki M. Life at the extreme limit: phenotypic characteristics of supercentenarians in Okinawa. J Gerontol A Biol Sci Med Sci. 2008;63(11):1201–8.
11. Takao M, Hirose N, Arai Y, Mihara B, Mimura M. Neuropathology of supercentenarians—four autopsy case studies. Acta Neuropathol Commun. 2016;4(1):97.
12. Lapin A, Böhmer F. Laboratory diagnosis and geriatrics: more than just reference intervals for the elderly. Wien Med Wochenschr. 2005;155(1–2):30–5.
13. Lio D, Malaguarnera M, Maugeri D, Ferlito L, Bennati E, Scola L, et al. Laboratory parameters in centenarians of Italian ancestry. Exp Gerontol. 2008;43(2):119–22.
14. Monti D, Ostan R, Borelli V, Castellani G, Franceschi C. Inflammaging and human longevity in the omics era. Mech Ageing Dev. 2017;165(Pt B):129–38.
15. Evert J, Lawler E, Bogan H, Perls T. Morbidity profiles of centenarians: survivors, delayers, and escapers. J Gerontol A Biol Sci Med Sci. 2003;58(3):232–7.
16. Engberg H, Oksuzyan A, Jeune B, Vaupel JW, Christensen K. Centenarians—a useful model for healthy aging? A 29-year follow-up of hospitalizations among 40,000 Danes born in 1905. Aging Cell. 2009;8(3):270–6.
17. Tedone E, Huang E, O'Hara R, Batten K, Ludlow AT, Lai TP, Arosio B, Mari D, Wright WE, Shay JW. Telomere length and telomerase activity in T cells are biomarkers of high-performing centenarians. Aging Cell. 2019;18(1):e12859. https://doi.org/10.1111/acel.12859.
18. White MC, Holman DM, Boehm JE, Peipins LA, Grossman M, Henley SJ. Age and cancer risk: a potentially modifiable relationship. Am J Prev Med. 2014;46(3 Suppl 1):S7–15.
19. Zinger A, Cho WC, Ben-Yehuda A. Cancer and aging—the inflammatory connection. Aging Dis. 2017;8:611–27.
20. Leonardi GC, Accardi G, Monastero R, Nicoletti F, Libra M. Ageing: from inflammation to cancer. Immun Ageing. 2018;15:1.
21. Siegel RL, Miller KD, Jemal A. Cancer statistics, 2015. CA Cancer J Clin. 2015;65(1):5–29.
22. United States Cancer Statistics [Internet]. Rate of new cancers in the United States [cited 2015]. https://gis.cdc.gov/Cancer/USCS/DataViz.html.
23. Stanta G, Campagner L, Cavallieri F, Giarelli L. Cancer of the oldest old. What we have learned from autopsy studies. Clin Geriatr Med. 1997;13(1):55–68.

24. Miyaishi O, Ando F, Matsuzawa K, Kanawa R, Isobe K. Cancer incidence in old age. Mech Ageing Dev. 2000;117(1–3):47–55.
25. Andersen SL, Terry DF, Wilcox MA, Babineau T, Malek K, Perls TT. Cancer in the oldest old. Mech Ageing Dev. 2005;126(2):263–7.
26. Carrasco-Garcia E, Moreno M, Moreno-Cugnon L, Matheu A. Increased Arf/p53 activity in stem cells, aging and cancer. Aging Cell. 2017;16(2):219–25.
27. Accardi G, Caruso C. Immune-inflammatory responses in the elderly: an update. Immun Ageing. 2018;15:11.
28. Gangemi S, Basile G, Monti D, Merendino RA, Di Pasquale G, Bisignano U, et al. Age-related modifications in circulating IL-15 levels in humans. Mediat Inflamm. 2005;2005(4):245–7.
29. Morrisette-Thomas V, Cohen AA, Fülöp T, Riesco É, Legault V, Li Q, et al. Inflamm-aging does not simply reflect increases in pro-inflammatory markers. Mech Ageing Dev. 2014;139: 49–57.
30. Gerli R, Monti D, Bistoni O, Mazzone AM, Peri G, Cossarizza A, et al. Chemokines, sTNF-Rs and sCD30 serum levels in healthy aged people and centenarians. Mech Ageing Dev. 2000;121(1–3):37–46.
31. Franceschi C, Capri M, Monti D, Giunta S, Olivieri F, Sevini F, et al. Inflammaging and anti-inflammaging: a systemic perspective on aging and longevity emerged from studies in humans. Mech Ageing Dev. 2007;128(1):92–105.
32. Franceschi C, Ostan R, Santoro A. Nutrition and inflammation: are centenarians similar to individuals on calorie-restricted diets? Annu Rev Nutr. 2018;38:329–56.
33. De la Fuente M, Miquel J. An update of the oxidation-inflammation theory of aging: the involvement of the immune system in oxi-inflamm-aging. Curr Pharm Des. 2009;15:3003–26.
34. Chandel NS. Mitochondrial regulation of oxygen sensing. Adv Exp Med Biol. 2010;661:339–54.
35. Bianchi ME. DAMPs, PAMPs and alarmins: all we need to know about danger. J Leukoc Biol. 2007;81(1):1–5.
36. Bullone M, Lavoie JP. The contribution of oxidative stress and inflamm-aging in human and equine asthma. Int J Mol Sci. 2017;18(12). pii: E2612.
37. Cui H, Kong Y, Zhang H. Oxidative stress, mitochondrial dysfunction, and aging. J Signal Transduct. 2012;2012:646354.
38. Fried LP, Tangen CM, Walston J, Newman AB, Hirsch C, Gottdiener J, et al. Frailty in older adults: evidence for a phenotype. J Gerontol A Biol Sci Med Sci. 2001;56:M146–56.
39. Paolisso G, Tagliamonte MR, Rizzo MR, Manzella D, Gambardella A, Varricchio M. Oxidative stress and advancing age: results in healthy centenarians. J Am Geriatr Soc. 1998;46:833–8.
40. Barbieri M, Rizzo MR, Manzella D, Grella R, Ragno E, Carbonella M, et al. Glucose regulation and oxidative stress in healthy centenarians. Exp Gerontol. 2003;38:137–43.
41. Mecocci P, Polidori MC, Troiano L, Cherubini A, Cecchetti R, Pini G, et al. Plasma antioxidants and longevity: a study on healthy centenarians. Free Radic Biol Med. 2000;28(8):1243–8. Erratum in: Free Radic Biol Med. 2000;29(5):486.
42. Suzuki M, Willcox DC, Rosenbaum MW, Willcox BJ. Oxidative stress and longevity in okinawa: an investigation of blood lipid peroxidation and tocopherol in okinawan centenarians. Curr Gerontol Geriatr Res. 2010;380460:1–10.
43. Nagase M, Yamamoto Y, Matsumoto N, Arai Y, Hirose N. Increased oxidative stress and coenzyme Q10 deficiency in centenarians. J Clin Biochem Nutr. 2018;63(2):129–36.
44. Conte M, Ostan R, Fabbri C, Santoro A, Guidarelli G, Vitale G, et al. Human aging and longevity are characterized by high levels of mitokines. J Gerontol A Biol Sci Med Sci. 2018 Jun 27. [Epub ahead of print].
45. Durieux J, Wolff S, Dillin A. The cell-non-autonomous nature of electron transport chain-mediated longevity. Cell. 2011;144(1):7–91.
46. Edrey YH, Salmon AB. Revisiting an age-old question regarding oxidative stress. Free Radic Biol Med. 2014;71:368–78.
47. Alzheimer's Disease International [Internet]. World Alzheimer Report 2018 [cited 2018 Sept]. https://www.alz.co.uk/research/world-report-2018.

48. Jopp DS, Park MK, Lehrfeld J, Paggi ME. Physical, cognitive, social and mental health in near-centenarians and centenarians living in New York City: findings from the Fordham Centenarian Study. BMC Geriatr. 2016;16:1.
49. Poon LW, Woodard JL, Stephen Miller L, Green R, Gearing M, Davey A, et al. Understanding dementia prevalence among centenarians. J Gerontol A Biol Sci Med Sci. 2012;67(4):358–65.
50. Arosio B, Ostan R, Mari D, Damanti S, Ronchetti F, Arcudi S, et al. Cognitive status in the oldest old and centenarians: a condition crucial for quality of life methodologically difficult to assess. Mech Ageing Dev. 2017;165:185–94.
51. Lescai F, Chiamenti AM, Codemo A, Pirazzini C, D'Agostino G, Ruaro C, et al. An APOE haplotype associated with decreased ε4 expression increases the risk of late onset Alzheimer's disease. J Alzheimers Dis. 2011;24:235–45.
52. Ishioka YL, Gondo Y, Fuku N, Inagaki H, Masui Y, Takayama M, et al. Effects of the APOE ε4 allele and education on cognitive function in Japanese centenarians. Age. 2016;38:495–503.
53. Kawasaki M, Arai Y, Takayama M, Hirata T, Takayama M, Abe Y, et al. Carotid atherosclerosis, cytomegalovirus infection, and cognitive decline in the very old: a community-based prospective cohort study. Age. 2016;38(2):29.
54. Nakagawa T, Cho J, Gondo Y, Martin P, Johnson MA, Poon LW, et al. Subjective well-being in centenarians: a comparison of Japan and the United States. Aging Ment Health. 2017;6:1–8.
55. MeSH Browser [Internet]. Analytical, diagnostic and therapeutic techniques and equipment category. Somatotypes. https://www.ncbi.nlm.nih.gov/mesh/68013008.
56. Khalil SF, Mohktar MS, Ibrahim F. The theory and fundamentals of bioimpedance analysis in clinical status monitoring and diagnosis of diseases. Sensors (Basel). 2014;14(6):10895–928.
57. Beleigoli AM, Boersma E, Diniz Mde F, Lima-Costa MF, Ribeiro AL. Overweight and class I obesity are associated with lower 10-year risk of mortality in Brazilian older adults: the Bambuí Cohort Study of Ageing. PLoS One. 2012;7(12):e52111.
58. Flegal KM, Kit BK, Orpana H, Graubard BI. Association of all-cause mortality with overweight and obesity using standard body mass index categories: a systematic review and meta-analysis. JAMA. 2013;309(1):71–82.
59. Kouvari M, Chrysohoou C, Tsiamis E, Kosyfa H, Kalogirou L, Filippou A, et al. The "overweight paradox" in the prognosis of acute coronary syndrome for patients with heart failure-A truth for all? A 10-year follow-up study. Maturitas. 2017;102:6–12.
60. Snijder MB, van Dam RM, Visser M, Seidell JC. What aspects of body fat are particularly hazardous and how do we measure them? Int J Epidemiol. 2006;35(1):83–92.
61. De Lorenzo A, Bianchi A, Maroni P, Iannarelli A, Di Daniele N, Iacopino L, et al. Adiposity rather than BMI determines metabolic risk. Int J Cardiol. 2013;166(1):111–7.
62. Cowen S, Hannan WJ, Ghosh S. Nutrition index determined by a portable multifrequency bioelectrical impedance analysis machine. GUT. 1998;42:144–52.
63. Guglielmi FW, Mastronuzzi T, Pietrini L, Panarese A, Panella C, Francavilla A. Electrical bioimpedance methods: applications to medicine and biotechnology. Ann N Y Acad Sci. 1999;873:105–11.
64. Gupta D, Lammersfeld CA, Burrows JL, Dahlk SL, Vashi PG, Grutsch JF, et al. Bioelectrical impedance phase angle in clinical practice: implications for prognosis in advanced colorectal cancer. Am J Clin Nutr. 2004;80(6):1634–8.
65. Barbosa-Silva MC, Barros AJ, Wang J, Heymsfield SB, Pierson RN Jr. Bioelectrical impedance analysis: population reference values for phase angle by age and sex. Am J Clin Nutr. 2005;82(1):49–52.

Centenarian Offspring as a Model of Successful Ageing

3

Anna Aiello, Mattia Emanuela Ligotti,
and Andrea Cossarizza

3.1 Introduction

The interest in how to age "successfully" and, in particular, the research of determinant factors for growing older in active way and in good health have increased in the last decade [1].

The concept of "successful ageing" (SA) has been used by gerontologists since the 1980s. Although there is no universally accepted definition of SA, because it varies among individuals, the World Health Organization defines it as "the process of developing and maintaining the functional ability that enables well-being in older age" [2]. Researchers have drawn up some criteria that describe SA [3]; these include three main related components: low probability to develop diseases and disease-related disabilities, high cognitive and physical functional capacity, and active engagement with life. So, SA can be considered more than the absence of disease or the maintenance of functional capacities because the combination of these components guarantees an active end of life that represents the essential concept of SA [4].

Scientists who study SA have been trying to determine which factors lead to a long, healthy life and design strategies that help to maintain health as we age. To do this, it is necessary to identify biological and molecular factors able to promote a healthy ageing by the deep study of a model.

A. Aiello (✉) · M. E. Ligotti
Laboratory of Immunopathology and Immunosenescence, Department of Biomedicine, Neurosciences and Advanced Diagnostics, University of Palermo, Corso Tukory 211, Palermo, Italy
e-mail: anna.aiello@unipa.it; mattiaemanuela.ligotti@unipa.it

A. Cossarizza
Department of Medical and Surgical Sciences of Children and Adult, University of Modena and Reggio Emilia, Modena, Italy
e-mail: andrea.cossarizza@unimore.it

© Springer Nature Switzerland AG 2019
C. Caruso (ed.), *Centenarians*, https://doi.org/10.1007/978-3-030-20762-5_3

Centenarians, i.e., subjects who live 100 or more years in good physical and mental conditions, represent the best model of SA [5]. They have reached the extreme limits of the lifespan overcoming the main age-related diseases. Several studies that define the phenotype of these exceptional subjects were carried on (see Chap. 2). However, these individuals have some restrictions due to frailty, which is a physiological characteristic of the extreme age, lack of an age-matched control group, and limited number of exponents [6].

In order to overcome some of these difficulties, several prototypes to study SA were searched. That of centenarian offspring (CO) has always been considered one of the best models to analyse healthy ageing and to study its determinants. In addition, CO are about 20–30 years younger than centenarians, and this facilitates the recruitment of their "natural" controls, namely age-matched subjects born from non-centenarian people [7].

CO share a familial trait principally influenced by genetic factors and environmental conditions [8]. In this regard, it was demonstrated that centenarians and, in general, long-living individuals (LLI) have a greater possibility to have offspring who live longer than offspring of non-long-living subjects [9]. In fact, human longevity seems to cluster in families enriched in long-living parents and ancestors, and parents who later will become centenarians likely adopt more healthy choices for their children [10]. Additionally, the research suggests that relatives (twin, sibling, and offspring) of LLI have a lower risk to experience the main age-related diseases, such as cardiovascular diseases (CVD), diabetes, and cancer. This is possibly associated with favourable gene equipment, lipid profile, healthy lifestyle, successful immunosenescence, among others [11] (see Fig. 3.1).

In this chapter, we summarize the current knowledge about CO phenotype in relation to understanding SA. However, it must be admitted that, despite the interest of this model to study determinants and trajectories of healthy ageing, a comprehensive and deep characterization of it is still lacking.

3.2 Hereditary and Modifiable Factors Influencing Longevity in Centenarian Offspring

Longevity and healthy ageing are among the most complex systems studied to date by gerontologists. These traits are heterogeneous, genetic [12], and epigenetic, and environmental factors contribute in different ways [13]. Nowadays, in adulthood, the heritability of age of death is approximately 25% [14], but it can reach even 48% in males who achieve exceptional longevity [15]. The remaining part is due to environmental exposures, chance, stochastic events, access to healthcare, and lack of trivial accidents [13]. Certainly, a very long life (over 95 years) has a strong genetic basis, and, indeed, several diseases that lead to death at an early age are due to the alteration of known genetic pathways [16–18]. Many of the genetic aspects of longevity include mutations or polymorphisms, which can occur at different frequencies within a population [19, 20].

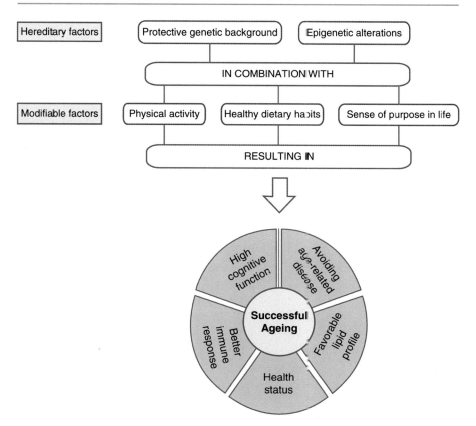

Fig. 3.1 A schematic overview of successful ageing. For explanation, see the text

Longevity genes affect a vast spectrum of biological functions that improve the feature of repair mechanisms, increase the resistance to stressors like virus and injury, and can slow the age-related senescent changes in cells and tissues.

These aspects are consistent with the data about CO, who have much less diabetes and CVD (see Sect. 3.4.2) than age-matched controls, suggesting that they have inherited a set of genes, from long-lived parents, that have protected them against age-related pathologies. In contrast, mutations in specific genes reduce lifespan by increasing the risk of a single fatal disease or a set of closely related diseases [21].

In this regard, many genetic studies, performed by the most advanced technologies, show contradictory results [20]. Linkage studies in exceptionally long-living families support the existence of a longevity locus in chromosome 3; other studies showed the existence of different loci likely linked to longevity. Among candidate genes assessed for their association with longevity, variants in APOE and FOXO3A have been most consistently confirmed, though some other genes have been associated with longevity phenotypes in some populations but not in others [19] (for the genetics of healthy ageing and longevity, see Chap. 6). In particular, genome-wide association scans of centenarians vs. young controls reveal that only APOE achieved statistical significance. However, analyses of combinations of single nuclear poly-morphisms (SNPs) or genes have identified pathways and signatures that converge upon biological processes associated with ageing development [14].

Concerning epigenetic factors, i.e., stable hereditary changes in gene expression that do not involve modifications in DNA sequence (i.e., a change in phenotype without a change in genotype), which, in turn, affect how cells read genes, they play an important role in several cellular activities (cellular senescence, tumorigenesis, and several age-related diseases) [22, 23]. However, few studies exist on their role on human longevity. Only in recent years, epigenomic alterations were increasingly recognized as a part of ageing, and they have been associated with the pathologic aspects of this phenotype (see Chap. 7).

In general, a significant decrease in global DNA methylation levels was observed with age. Gentilini et al. demonstrated that CO have a better health status, and interestingly, age-related loss of DNA methylation was less pronounced [24]. In order to test whether families with extreme longevity are epigenetically distinct from others, according to an epigenetic biomarker of ageing that is known as "epi-genetic clock", some authors analysed DNA methylation levels of peripheral blood mononuclear cells from Italian families. These families were formed by semi-supercentenarians (mean age: 105.6 ± 1.6 years), semi-supercentenarian offspring (mean age: 71.8 ± 7.8 years), and age-matched controls (mean age: 69.8 ± 7.2 years). They demonstrated that the offspring of semi-supercentenarians have a lower epi-genetic age than age-matched controls (age difference = 5.1 years, $p = 0.00043$) and that centenarians are younger (8.6 years) than expected based on their chrono-logical age [25].

Other nongenetic or epigenetic factors, particularly environment and lifestyle, affect the development of age-related diseases and the healthy span in the general population [26]. The correlation between genetic background and environment in determining the individual chance of a delayed or SA is a hot topic in gerontology. The exact knowledge of the effects of environment and lifestyle on the basic molec-ular mechanisms of ageing may be used as preventive measures to increase the chance to attain longevity. Among many aspects of environment that can affect the maintenance of a healthy ageing phenotype, physical activity and healthy dietary habits are the most important modifiable factors [27]. In particular, nutrition is a daily process by which living organisms get food to gather energy and nutrients to live. Good nutrition plays a significant role in determining the well-being of older people and in delaying and reducing the risk of contracting diseases. Nutritionally unbalanced diets are often associated with diabetes and with the risks of developing coronary heart diseases. It was seen that children who follow a balanced diet with lots of fruits and vegetables are likely to continue eating healthily into adulthood. On the contrary, parents who eat too much processed food with high levels of salt, sugar, and fat tend to pass those habits onto their own children [28].

The maintenance of physical function is a key component of most definitions of SA [29]. An interesting study demonstrated that persons with long-lived parents might enjoy a longer life spared from physical functioning decline [30]. These results are supported by a previous analysis of three Danish population-based studies that reported that parental longevity was associated with better physical and cognitive function measures in the adult offspring [31].

Concluding, the existence proof of the correlation between "healthy lifestyle and environments" comes also from the observation of places, called "Blue Zones", where the population shares a common lifestyle in a common and geographically defined environment (see Chap. 4).

3.3   Phenotypic Features of Centenarian Offspring

As previously stated, CO represent a suitable model to study some age-dependent phenotypic variables potentially involved in the modulation of the lifespan. Indeed, the comparison of CO with their controls has already been successfully applied to identify biochemical and metabolomics parameters related to exceptional longevity [32–38]. However, despite the interest of the CO model to study determinants and trajectories of SA, a comprehensive and deep characterization of the CO phenotype is still lacking.

In some cross-sectional and longitudinal studies, it was seen that weight gain, increased body mass index (BMI), and obesity contribute to a decline in physical and cognitive activities; an increased body fat, particularly visceral fat and high BMI (see Chap. 2), has been associated with insulin resistance, higher risk for diabetes, cardiovascular disease, hypertension, and metabolic disorders [39–42].

In 2016, Bucci et al. conducted in Italy the first comprehensive investigation of the health status of CO compared to age-matched offspring of non-LLI (non-centenarian offspring, NCO). From this multidisciplinary approach, it emerged that CO showed a better physical status and a lower prevalence of pathological conditions than NCO, as reported in previous studies with other populations [43–47].

In addition, they had a higher education level and a lower proportion of married individuals than NCO, although no reasoning was provided. They appeared leaner than NCO, and indeed, they had lower weight, smaller waist and hip circumferences, and lower BMI. However, these anthropometric features show a gender distinction because these significant differences were evident only in female groups. Also regarding the brawn, CO showed higher scores in the handgrip test compared to controls [47].

In another study, in order to clarify the role of the insulin growth factor (IGF)-1 system in human lifespan, circulating IGF-1 bioactivity and other parameters of the IGF/insulin system were measured in CO and in a properly selected group of offspring-matched controls. The IGF-1/insulin pathway is the main focus of the ageing and longevity research for the association of its SNPs with SA and longevity. In human beings, ageing is associated with lower IGF-1 circulating levels [48], and in long-lived people, IGF-1 receptor (IGF-1R) has been correlated with modulation of human lifespan through the attenuation of IGF-1 signalling [49]. Both IGF-1 and IGF-1R polymorphisms that theoretically can modulate IGF-1 pathway have been studied for their correlation with longevity, but evidences to date are not conclusive [49–53]. Circulating IGF-1 bioactivity and total IGF-1 were significantly lower in

CO than in offspring-matched controls (both $p < 0.01$) [33]. In the same study, low IGF-1 bioactivity in CO was inversely associated with insulin sensitivity, confirming previous data described in elderly [54–56] and healthy subjects. This relationship remained after corrections for age, gender, and BMI. In addition, centenarians showed a twofold higher insulin sensitivity and comparable glucose levels than their younger offspring, despite a lower circulating IGF-1 bioactivity [57, 58].

The preservation of cognitive fitness such as a good processing speed, learning, and memory is an important goal to SA because fitness significantly affects the quality of life. Lifestyle and pharmacological interventions can ameliorate the cognitive function, but literature data indicate the importance of a familial component, like inheritance of genetic factors, to protect against cognitive impairment. More specifically, CO, especially females, have a higher standardized mini-mental state examination scores, a lower risk of Alzheimer's disease [59], a reduced incidence of cognitive decline, and, overall, a significantly better cognitive fitness in comparison with age-matched controls [47, 60]. A case–control study was conducted to determine whether the CO have personality characteristics that are distinct from the general population [61]. Because some personality traits have substantial heritable components, it was hypothesized that certain personality features may be important to reach healthy ageing. CO were characterized by lower neuroticism, and females revealed greater agreeableness [62].

The sense of purpose in life, defined as "a feeling of meaning and direction in one's life" [63], has received increasing attention in the last years because of its association with positive health outcomes, including healthy cognitive ageing. In this context, a recent research found that CO have significantly higher purpose in life scores than controls, suggesting that this condition of psychological well-being may be contributing to their SA profiles [64]. Therefore, measures of personality might be an important phenotype to include in studies that assess genetic and environmental influences of longevity and SA.

3.4 Age-Related Disease Risk Profile in Centenarian Offspring

The increase in human life expectancy, occurred during the last years, is certainly a public health success, but it is also the main risk factor for the onset of chronic diseases, including cancer, CVD, and neurodegenerative diseases, which are characteristic of ageing.

As stated in the previous paragraph, several studies from different populations are concordant that the CO show a significantly healthier status, with a reduced prevalence of pathological conditions than the elderly, proving that staying relatively healthy is physiologically possible in old age [7].

In this regard, to evaluate the relative incidence of age-related diseases, CO and referent cohort subjects (spouses of CO) were enrolled in a longitudinal study in order to discern whether the differences in the development of age-related diseases, between CO and the control group, persist over the follow-up period or whether the differences disappear as the two groups get older. Questionnaires about sociodemographic factors and history of the main age-related diseases were administered to the enrolled subjects. Additionally, participant physician carried out a routine physical examination for height, weight, blood pressure, and temperature at a routine examination. It was seen that CO were less likely to die during the period of follow-up evaluation than were equivalent control group. Moreover, CO had 81% lower odds of mortality during the period between the administration of the health questionnaire and the follow-up. However, CO and the referent cohort had similar levels of functional decline [65].

Finally, assuming that poor oral health has been associated with age-related diseases, an observational cross-sectional study of centenarians and their offspring was conducted in order to clarify if a better oral health is related to longevity and healthy span. These subjects were compared with referent cohort participants from the New England Centenarian Study [66]. The referent cohort, in comparison to the CO, was more likely to be edentulous, less likely to report excellent/very good oral health, and less likely to have all or more than half of their own teeth, suggesting that CO exhibit better oral health than their respective birth cohorts [66].

3.4.1 Inflammatory Status

In a sizable longitudinal study of semi-supercentenarians, conducted by Japanese groups, to determine the most important forces of SA at exceptional longevity, data from multiple studies were collected concerning haematopoiesis, inflammation, lipid, and glucose metabolism, among others. Inflammation (for 10.8%) and cognition (8.6%) predicted all-cause mortality in very old people better than chronologic age or gender. Moreover, the inflammation score was also lower in CO compared with age-matched controls. Furthermore, centenarians and their children maintain long telomeres despite the fact that telomere length was not considered a predictor of SA in semi-supercentenarian and centenarian people [34]. Therefore, the authors conclude that inflammation is an important driver of ageing and markers of inflammation predicted possible morbidities. These data confirm the theories that considered the inflammation one of the main causes of age-related diseases [67].

3.4.2 CVD Risk Profile

As regards the cardiovascular field, several studies agree that CO, like their parents, have lower prevalence and incidence of hypertension, myocardial infarction, and strokes. A growing number of evidence suggests that a combination of genetic factors and lifestyle aspects allows centenarians and their offspring to overcome cardiovascular events. The importance of genetic contribution is demonstrated by the inheritance of low risk for CVD in CO compared with their spouses or with NCO. In the same way, it seems that CO delay the onset of CVD with respect to age- and sex-matched control groups [68]. In addition, measurements of body fat distribution show a lower prevalence of abdominal obesity, one of the major risk factors for CVD, in CO than in controls [32].

As previously stated, the related question about the significant survival advantage in families with exceptional longevity, namely, if it is due to lifestyle and dietary patterns, to genetic factors, or both, suggested to perform a survey in the Ashkenazi population living in the United States. This study compared CO and controls with similar lifestyle (physical activity, smoking, and alcohol use), socioeconomic status, and dietary intake. In line with other studies, CO show a considerable reduction in the risk of hypertension, CVD, and stroke, thus highlighting the potential contribution of genetics in SA among CO [69].

Nevertheless, in a study conducted in 2016, Italian CO seem to show a worse lipid profile than controls, with higher levels of total and LDL cholesterol and lower levels of HDL cholesterol, but always within the normal range [47]. This apparent negative total/HDL cholesterol ratio can be explained by the fact that a lower number of Italian CO recruited, in comparison to their controls, use lipid-lowering drugs.

3.5 Other Biomarkers of Successful Ageing in Centenarian Offspring

Other possible biomarkers had been proposed by the researches in the last decade.

One of this was the heat shock protein (HSP) 70. This protein protects exposed cells from different types of stress. On the contrary, the extracellular variant has been shown to have both protective and damaging roles. Some evidences have suggested that the higher circulating levels of HSP70 can predict the future development of CVD. In this regard, Terry et al. analysed serum level of HSP70 in centenarians and their relatives. CO showed approximately tenfold lower levels of circulating serum HSP70 compared with spousal controls. The authors hypothesized that circulating HSP70 was correlated with diseases or disorders in which there is destruction or a damage to target tissues or organs, including CVD and autoimmune disorders [44]. Successively, the same authors suggested that low levels of circulating serum HSP70 could be an indicator of a healthy state and exceptional longevity. In order to assess the effect of HSP70 on human longevity, the research group conducted a pilot cross-sectional study, recruiting centenarians and CO. They analysed two polymorphisms of the HSP70 gene and the related serum level and established that no genetic association was found, but centenarians and their offspring had low HSP70 serum level than unrelated controls [70].

Researches have focused their attention on the effects of microRNAs (miRNAs) on the inflammatory status in ageing process. The role of circulating miRNAs on important cellular processes is well understood, but little is known about their expression throughout the ageing process. In particular, it was seen that miR-21 expression was higher in the CVD subjects and lower in the CO compared to that in the age-matched NCO. The authors suggested that circulating miR-21 might be an inflammatory biomarker linking the ageing process to age-related diseases like CVD. Indeed, previous findings showed the correlation between the expression of some markers of inflammation, such as interleukin-6, tumour necrosis factor-α, and C reactive protein, and miR-21, and the development of disabilities in older subjects [71].

Table 3.1 Overview of factors associated with successful ageing in CO

Variables	Comparison to NCO
Epigenetic factors	
DNA methylation	Less pronounced age-related loss of DNA methylation
Phenotypic features	
Physical status	Lower weight, smaller waist and hip circumferences, lower BMI and healthier status
Handgrip test	Higher scores
IGF-1 bioactivity	Significantly lower ($p < 0.01$)
Oral health	Excellent/very good
Personality traits	Higher purpose in life scores and lower neuroticism
Age-related disease risk profile	
Mortality	Lower odds ($p < 0.01$)
Alzheimer's disease	Lower risk
Cognitive function	Higher mini-mental state scores, lower incidence of cognitive decline and better cognitive fitness
Inflammation	Lower score
CVD	Lower prevalence and incidence of hypertension, myocardial infarction and strokes; lower serum HSP70 levels and miR-21 expression

NCO non-centenarian offspring, *BMI* body mass index, *IGF-1* insulin growth factor-1, *CVD* cardiovascular diseases, *HSP70* heat shock protein 70, *miR-21* microRNA-21, *p* p-value

Table 3.1 reports an overview of main characteristic features of CO.

3.6 Immunophenotype of Centenarian Offspring and Successful Immunosenescence

In 1969, Roy Walford published his landmark book, "The Immunologic Theory of Aging" [72]. Briefly, he hypothesized that the normal process of ageing in man and in all animals is pathogenetically related to faulty immune processes. Therefore, he was the first to note and promote the power of modern immunological approaches as tools for the analysis of ageing.

In fact, many alterations in innate and acquired immunity have been described in the elderly and considered deleterious, hence the term immunosenescence. On the other hand, immunosenescence is a complex process involving multiple reorganizational and developmentally regulated changes, rather than simple unidirectional decline of the whole function. However, some immunological parameters are commonly notably reduced in the elderly, and reciprocally, good function is strictly correlated with health status [73].

As discussed in this book, centenarians show a complex and heterogeneous phenotype, which seems to be the result of the capacity to adapt and remodel their body in response to stressors that makes them able to delay the ageing process and to escape the major age-related diseases. Accordingly, centenarian immune system shares characteristics both of young and elderly people. Number and function of NK cells (particularly, NK CD56dimCD16$^+$ characterized by increased cytotoxicity) are well conserved and comparable with those observed in young, whereas T and B cell number and function are similar to those observed in the elderly (although T cells show an increasing trend when compared to the elderly) [74–76]. However, a limitation of these studies is the lack of appropriate controls. As discussed in this chapter, CO have a significant advantage for longer survival and can resolve the problems occurring with their parents because they can be compared with an appropriate control group, consisting of age-matched elderly people whose parents died at an average life expectancy.

Ageing is characterized by the decline or as generally recognized by the remodelling of the immune system, with several modifications that include a reduced amount of T, B, and NK lymphocytes. In the elderly, a chronic antigenic load causes the filling of the immunological space by a population of CD8$^+$ T lymphocytes with a late-differentiated phenotype and the shrinkage of the T cell repertoire [77]. Indeed, due to lifelong and chronic exposure to pathogens, as in the case of persistent infection, T cells replicate several times and further differentiate in late-differentiated effector memory CD8$^+$ T cells with features of replicative senescence [78, 79], an irreversible growth arrest, becoming significantly less able to commit in efficient immune responses. However, senescent T lymphocytes can remain metabolically active for a prolonged period of time and adopt a secretory profile, with secretion of pro-inflammatory cytokines resulting in a low-grade chronic inflammatory state known as "inflammaging" which is strongly associated with many age-related diseases (see above) [80].

The impairment of circulating T cell compartment together with the reduced output of newly formed T cells from the thymus in elderly subjects lead to an inadequate immune response against newly encountered antigens. The apparently inevitable consequence of this complex scenario is the reduction of the availability of the elderly to respond to novel antigens and to vaccines, resulting in an increased susceptibility to infection and development of age-related disorders [81–83].

In these last years, some researchers have speculated about the distinctive immunological profile of offspring enriched for longevity with respect to the immunological features of coeval elderly. The human cytomegalovirus (HCMV) is one of the most common viruses that affect elderly people. Many evidences have shown that HCMV infection may influence the T cell subset distribution, having an essential role in immunosenescence [84, 85]. CMV infection is strongly related to both a reduction of naïve (CD45RA$^+$CCR7$^+$CD27$^+$CD28$^+$) CD8$^+$ T cells and a contemporary increase in late-differentiated effector memory (T$_{EM}$, CD8$^+$CD45RA$^-$CCR7$^-$CD27$^-$CD28$^-$) and effector memory T cells reexpressing the naïve cell marker CD45RA (T$_{EMRA}$, CD8$^+$CD45RA$^+$CCR7$^-$CD27$^-$CD28$^-$). These parameters are considered typical of immunosenescence in the elderly. HCMV-seropositive offspring of long-lived people do not show the age-associated decrease of naïve T cells. On the other hand, memory T cell subsets described above do not increase in offspring of long-lived families, different from that observed in age-matched controls.

It has also been demonstrated that HCMV-seropositive offspring of long-lived people have reduced levels of CD8$^+$ T cells expressing CD57 and KLRG1, sometimes referred as "marker of replicative senescence", when compared with their CMV-infected age-matched controls. The reduction of T$_{EM}$ and T$_{EMRA}$ cells with features of replicative senescence observed in HCMV-infected offspring could explain their high proliferative response against CMV [86].

As previously stated, ageing also has a strong impact on the remodelling of the B cell compartment. An evident effect is the significant decrease of circulating B cells, mainly due to the reduction of new B cell produced in the bone marrow. Furthermore, in the elderly, there is a shift from naïve (IgD, IgM) to memory (IgG, IgA) immunoglobulin production, accompanied by the impaired ability to produce high-affinity protective antibodies against newly encountered pathogens and vaccines. Finally, in elderly subjects, a reduction of naïve B IgD$^+$CD27$^-$ cells exists along with an increase of a B cell subpopulation with a senescence-associated phenotype, IgD$^-$CD27$^-$ Double Negative (DN) late memory B cells [87–91].

Concerning the B cell compartment in CO, they do not present the typical naïve-memory shift observed in the elderly, although they present a reduction in the B cell count, typical of old people. Moreover, CO do not show neither the increase of the IgD$^-$CD27$^-$ DN late memory B cells nor the decrease of IgD$^+$CD27$^-$ naïve B cells, observed in elderly people, but they look more similar to young people. The failure of age-associated increase of DN B cells, as observed in the general elderly population, suggests that in CO there is no exhaustion of the B cell compartment. Furthermore, the evaluation of CO serum IgM values shows that these values are within the range of the levels displayed by young individuals. All [88, 91–93] these data suggest a good bone marrow cell reservoir in CO, despite the fact that the ability of bone marrow to generate B cells is impaired with age.

Therefore, both T [94] and B compartments of CO do not show the typical feature of immune system ageing, i.e., the shift from naïve to exhausted memory cells, but a "younger" immune profile.

See Table 3.2 for the immunological profile overview.

Table 3.2 Cellular age-related modification of T and B cell compartments in centenarian offspring compared to the elderly

Cell subset	Phenotype	Modification in CO	Modification in elderly
Naïve T CD8$^+$	CD3$^+$/CD8$^+$/CD45RA$^+$/CCR7$^+$/CD27$^+$/CD28$^+$	Increase	Decrease
T$_{EM}$	CD3$^+$/CD8$^+$/CD45RA$^-$/CCR7$^-$/CD27$^-$/CD28$^-$	Decrease	Increase
T$_{EMRA}$	CD3$^+$/CD8$^+$/CD45RA$^+$/CCR7$^-$/CD27$^-$/CD28$^-$	Decrease	Increase
Naïve B	CD19$^+$/IgD$^+$/CD27$^-$	Increase	Decrease
DN B	CD19$^+$/IgD$^-$/CD27$^-$	Decrease	Increase

CO centenarian's offspring, *TEM* effector-memory T cell, *TEMRA* effector memory RA T cell, *DN* double negative

3.7   Conclusion

According to the findings of some studies, CO appear the most promising model for research on the determinant factors of longevity and SA because they have a familial trait. They show a favourable lipid, immunological, and cardiovascular profile, a decreased cognitive decline, and a protective genetic background. Moreover, CO present an immunological profile comparable for several aspects to that of young individuals. They maintain the ability to respond both to new natural infections and to vaccination, different from the elderly without centenarian parents and centenarians themselves.

The real strength of the studies that use this model to identify genetic, phenotypic, or metabolic strategic variants of healthy ageing is the possibility to compare CO with age-matched offspring of non-long-lived individuals, more easily recruited and available than centenarian itself.

The approach of this study has allowed and will continue to permit the identification of other targets or biomarkers of SA. Furthermore, it might consent to develop or identify anti-ageing therapies, modulating ageing rate and developing drugs (i.e., antioxidants) or new lifestyle habits (i.e., a healthy diet, caloric uptake reduction, use of prebiotics and probiotics, and increased physical exercise), to become older in an active and independent way.

In conclusion, the healthcare costs in many countries are very high because of the increased number of unhealthy populations and the consequent increase of severe age-related disabilities. However, experiments in laboratory organisms have shown that ageing is not an immutable process. Indeed, interventions to slow or postpone ageing and to increase the expectancy of an active life are available. Pharmacological therapy is one of these. Healthy nutrition could be another alternative, and the study of CO lifestyle can represent the possibility to change, in part, the course of the ageing process.

However, a note of caution should be addressed since the number of centenarians is also increasing because of improvements in hygienic and sanitary conditions and in the quality of nutrition that allow a greater number of older old to become exceptional. Thus, the current centenarians and even the future ones, as well as their offspring, will be likely much less selected than the centenarians of a decade ago.

References

1. Longo VD, Antebi A, Bartke A, Barzilai N, Brown-Borg HM, Caruso C, et al. Interventions to slow aging in humans: are we ready? Aging Cell. 2015;14:497–510.
2. Wykle M, Whitehouse P, Morris D. Successful aging through the life span: intergenerational issues in health. New York: Springer; 2005.
3. Montross LP, Depp C, Daly J, Reichstadt J, Golshan S, Moore D, et al. Correlates of self-rated successful aging among community-dwelling older adults. Am J Geriatr Psychiatry. 2006;14(1):43–51.
4. Bülow MH, Söderqvist T. Successful ageing: a historical overview and critical analysis of a successful concept. J Aging Stud. 2014;31:139–49.

5. Avery P, Barzilai N, Benetos A, Bilianou H, Capri M, Caruso C, et al. Ageing, longevity, exceptional longevity and related genetic and non genetics markers: panel statement. Curr Vasc Pharmacol. 2014;12:659–61.
6. Pyrkov TV, Getmantsev E, Zhurov B, Avchaciov K, Pyatnitskiy M, Menshikov L, et al. Quantitative characterization of biological age and frailty based on locomotor activity records. Aging. 2018;10(10):2973–90.
7. Gueresi P, Miglio R, Monti D, Mari D, Sansoni P, Caruso C, et al. Does the longevity of one or both parents influence the health status of their offspring? Exp Gerontol. 2013;48(4): 395–400.
8. Balistreri CR, Candore G, Accardi G, Buffa S, Bulati M, Martorana A, et al. Centenarian offspring: a model for understanding longevity. Curr Vasc Pharmacol. 2014;12(5):718–25.
9. Gudmundsson H, Gudbjartsson DF, Frigge M, Gulcher JR, Stefánsson K. Inheritance of human longevity in Iceland. Eur J Hum Genet. 2000;8(10):743–9.
10. Caselli G, Lapucci E, Lipsi RM, Pozzi L, Baggio G, Carru C, et al. Maternal longevity is associated with lower infant mortality. Demogr Res. 2014;31:1275–96.
11. Mansur Ade P, Mattar AP, Rolim AL, Yoshi FR, Marin JF, César LA, et al. Distribution of risk factors in parents and siblings of patients with early coronary artery disease. Arq Bras Cardiol. 2003;80(6):582–4, 579–81. Epub 2003 Jul 2. English, Portuguese. PubMed PMID: 12856068.
12. Herskind AM, McGue M, Holm NV, Sørensen TI, Harvald B, Vaupel JW. The heritability of human longevity: a population-based study of 2872 Danish twin pairs born 1870-1900. Hum Genet. 1996;97(3):319–23.
13. Passarino G, De Rango F, Montesanto A. Human longevity: genetics or lifestyle? It takes two to tango. Immun Ageing. 2016;13:12. https://doi.org/10.1186/s12979-016-0066-z. eCollection 2016. PubMed PMID: 27053941; PubMed Central PMCID: PMC4822264.
14. Brooks-WilsonAR. Genetics of healthy aging and longevity. Hum Genet. 2013;132(12):1323–38.
15. Puca AA, Spinelli C, Accardi G, Villa F, Caruso C. Centenarians as a model to discover genetic and epigenetic signatures of healthy ageing. Mech Ageing Dev. 2018;174:95–102.
16. Perls TT, Wilmoth J, Levenson R, Drinkwater M, Cohen M, Bogan H, et al. Life-long sustained mortality advantage of siblings of centenarians. Proc Natl Acad Sci U S A. 2002;99(12): 8442–7.
17. Martin GM, Oshima J, Gray MD, Poot M. What geriatricians should know about the Werner syndrome. J Am Geriatr Soc. 1999;47(9):1136–44.
18. Fossel M. Human aging and progeria. J Pediatr Endocrinol Metab. 2000;13(Suppl 6):1477–81.
19. Shadyab AH, LaCroix AZ. Genetic factors associated with longevity: a review of recent findings. Ageing Res Rev. 2015;19:1–7. Epub 2014 Nov 5.
20. Santos-Lozano A, Santamarina A, Pareja-Galeano H, Sanchis-Gomar F, Fiuza-Luces C, Cristi-Montero C, et al. The genetics of exceptional longevity: insights from centenarians. Maturitas. 2016;90:49–57. Epub 2016 May 10.
21. Yashin AI, Wu D, Arbeeva LS, Arbeev KG, Kulminski AM, Akushevich I, et al. Genetics of aging, health, and survival: dynamic regulation of human longevity related traits. Front Genet. 2015;6:122.
22. Kim M, Long TI, Arakawa K, Wang R, Yu MC, Laird PW. DNA methylation as a biomarker for cardiovascular disease risk. PLoS One. 2010;5(3):e9692
23. Rakyan VK, Down TA, Maslau S, Andrew T, Yang TP, Beyan H, et al. Human aging associated DNA hypermethylation occurs preferentially at bivalent chromatin domains. Genome Res. 2010;20:434–9.
24. Gentilini D, Mari D, Castaldi D, Remondini D, Ogliari G, Ostan R, et al. Role of epigenetics in human aging and longevity: genome-wide DNA methylation profile in centenarians and centenarians' offspring. Age. 2013;35(5):1961–73.
25. Horvath S, Pirazzini C, Bacalini MG, Gentilini D, Di Blasio AM, Delledonne M, et al. Decreased epigenetic age of PBMCs from Italian semi-supercentenarians and their offspring. Aging. 2015;7(12):1159–70.
26. Govindaraju D, Atzmon G, Barzilai N. Genetics, lifestyle and longevity: lessons from centenarians. Appl Transl Genom. 2015;4:23–32.

27. Dato S, Crocco P, D'Aquila P, de Rango F, Bellizzi D, Rose G, et al. Exploring the role of genetic variability and lifestyle in oxidative stress response for healthy aging and longevity. Int J Mol Sci. 2013;14:16443–72.
28. Kiefte-de Jong JC, Mathers JC, Franco OH. Nutrition and healthy ageing: the key ingredients. Proc Nutr Soc. 2014;73:249–59.
29. Peel NM, McClure RJ, Bartlett HP. Behavioral determinants of healthy aging. Am J Prev Med. 2005;28(3):298–304.
30. Ayers E, Barzilai N, Crandall JP, Milman S, Verghese J. Association of exceptional parental longevity and physical function in aging. Age. 2014;36(4):9677.
31. Frederiksen H, McGue M, Jeune B, Gaist D, Nybo H, Skytthe A, et al. Do children of long-lived parents age more successfully? Epidemiology. 2002;13(3):334–9.
32. Ostan R, Bucci L, Cevenini E, Palmas MG, Pini E, Scurti M, et al. Metabolic syndrome in the offspring of centenarians: focus on prevalence, components, and adipokines. Age. 2013;35(5):1995–2007. Epub 2012 Nov 9.
33. Vitale G, Brugts MP, Ogliari G, Castaldi D, Fatti LM, Varewijck AJ, et al. Low circulating IGF-I bioactivity is associated with human longevity: findings in centenarians' offspring. Aging. 2012;4(9):580–9.
34. Arai Y, Martin-Ruiz CM, Takayama M, Abe Y, Takebayashi T, Koyasu S, et al. Inflammation, but not telomere length, predicts successful ageing at extreme old age: a longitudinal study of semisupercentenarians. EBioMedicine. 2015;2(10):1549–58.
35. Fagnoni FF, Vescovini R, Passeri G, Bologna G, Pedrazzoni M, Lavagetto G, et al. Shortage of circulating naive CD8+ T cells provides new insights on immunodeficiency in aging. Blood. 2000;95(9):2860–8.
36. Franceschi C, Bonafè M, Valensin S, Olivieri F, De Luca M, Ottaviani E, et al. Inflamm-aging. An evolutionary perspective on immunosenescence. Ann N Y Acad Sci. 2000;908:244–54.
37. Larbi A, Franceschi C, Mazzatti D, Solana R, Wikby A, Pawelec G. Aging of the immune system as a prognostic factor for human longevity. Physiology. 2008;23:64–74.
38. De Benedictis G, Franceschi C. The unusual genetics of human longevity. Sci Aging Knowledge Environ. 2006;(10):pe20.
39. Paolisso G, Barbieri M, Bonafè M, Franceschi C. Metabolic age modelling: the lesson from centenarians. Eur J Clin Invest. 2000;30(10):888–94.
40. Dwimartutie N, Setiati S, Oemardi M. The correlation between body fat distribution and insulin resistance in elderly. Acta Med Indones. 2010;42(2):66–73.
41. Katzmarzyk PT, Reeder BA, Elliott S, Joffres MR, Pahwa P, Raine KD, et al. Body mass index and risk of cardiovascular disease, cancer and all-cause mortality. Can J Public Health. 2012;103(2):147–51.
42. Hwang LC, Bai CH, Sun CA, Chen CJ. Prevalence of metabolically healthy obesity and its impacts on incidences of hypertension, diabetes and the metabolic syndrome in Taiwan. Asia Pac J Clin Nutr. 2012;21(2):227–33.
43. Terry DF, Wilcox M, McCormick MA, Lawler E, Perls TT. Cardiovascular advantages among the offspring of centenarians. J Gerontol A Biol Sci Med Sci. 2003;58:M425–31. Erratum in: J Gerontol A Biol Sci Med Sci. 2008;63:706.
44. Terry DF, Wilcox MA, McCormick MA, Perls TT. Cardiovascular disease delay in centenarian offspring. J Gerontol A Biol Sci Med Sci. 2004;59(4):385–9.
45. Terry DF, Wilcox MA, McCormick MA, Pennington JY, Schoenhofen EA, Andersen SL, et al. Lower all-cause, cardiovascular, and cancer mortality in centenarians' offspring. J Am Geriatr Soc. 2004;52(12):2074–6.
46. Atzmon G, Schechter C, Greiner W, Davidson D, Rennert G, Barzilai N. Clinical phenotype of families with longevity. J Am Geriatr Soc. 2004;52(2):274–7.
47. Bucci L, Ostan R, Cevenini E, Pini E, Scurti M, Vitale G, et al. Centenarians' offspring as a model of healthy aging: a reappraisal of the data on Italian subjects and a comprehensive overview. Aging. 2016;8(3):510–9.
48. Bartke A. Minireview: role of the growth hormone/insulin-like growth factor system in mammalian aging. Endocrinology. 2005;146:3718–23.

49. Suh Y, Atzmon G, Cho MO, Hwang D, Liu B, Leahy DJ, et al. Functionally significant insulin-like growth factor I receptor mutations in centenarians. Proc Natl Acad Sci. 2008;105:3438–42.
50. Xie L, Gong YY, Lian SG, Yang J, Yang Y, Gao SJ, et al. Absence of association between SNPs in the promoter region of the insulin-like growth factor 1 (IGF-1) gene and longevity in the Han Chinese population. Exp Gerontol. 2008;43:962–5.
51. Bonafè M, Barbieri M, Marchegiani F, Olivieri F, Ragno E, Giampieri C, et al. Polymorphic variants of insulin-like growth factor I (IGF-I) receptor and phosphoinositide 3-kinase genes affect IGF-I plasma levels and human longevity: cues for an evolutionarily conserved mechanism of life span control. J Clin Endocrinol Metab. 2003;88:3299–304.
52. Albani D, Batelli S, Polito L, Vittori A, Pesaresi M, Gajo GB, et al. A polymorphic variant of the insulin-like growth factor 1 (IGF-1) receptor correlates with male longevity in the Italian population: a genetic study and evaluation of circulating IGF-1 from the "Treviso Longeva (TRELONG)" study. BMC Geriatr. 2009;9:19.
53. Barbieri M, Boccardi V, Esposito A, Papa M, Vestini F, Rizzo MR, et al. A/ASP/VAL allele combination of IGF1R, IRS2, and UCP2 genes is associated with better metabolic profile, preserved energy expenditure parameters, and low mortality rate in longevity. Age. 2012;34:235–45.
54. Brugts MP, van Duijn CM, Hofland LJ, Witteman JC, Lamberts SW, Janssen JA. Igf-I bioactivity in an elderly population: relation to insulin sensitivity, insulin levels, and the metabolic syndrome. Diabetes. 2010;59(2):505–8.
55. Arafat AM, Weickert MO, Frystyk J, Spranger J, Schöfl C, Möhlig M, et al. The role of insulin-like growth factor (IGF) binding protein-2 in the insulin-mediated decrease in IGF-I bioactivity. J Clin Endocrinol Metab. 2009;94(12):5093–101. Epub 2009 Oct 21.
56. Franceschi C, Bonafè M. Centenarians as a model for healthy aging. Biochem Soc Trans. 2003;31(2):457–61.
57. Paolisso G, Gambardella A, Ammendola S, D'Amore A, Balbi V, Varricchio M, et al. Glucose tolerance and insulin action in healthy centenarians. Am J Physiol. 1999;270:E890–4.
58. Barbieri M, Rizzo MR, Manzella D, Paolisso G. Age-related insulin resistance: is it an obligatory finding? The lesson from healthy centenarians. Diabetes Metab Res Rev. 2001;17(1):19–26.
59. Lipton RB, Hirsch J, Katz MJ, Wang C, Sanders AE, Verghese J, et al. Exceptional parental longevity associated with lower risk of Alzheimer's disease and memory decline. J Am Geriatr Soc. 2010;58(6):1043–9.
60. Andersen SL, Sweigart B, Sebastiani P, Drury J, Sidlowski S, Perls TT. Reduced prevalence and incidence of cognitive impairment among centenarian offspring. J Gerontol A Biol Sci Med Sci. 2019;74(1):108–13. https://doi.org/10.1093/gerona/gly141.
61. Givens JL, Frederick M, Silverman L, Anderson S, Senville J, Silver M, et al. Personality traits of centenarians' offspring. J Am Geriatr Soc. 2009;57(4):683–5.
62. Lewis NA, Turiano NA, Payne BR, Hill PL. Purpose in life and cognitive functioning in adulthood. Neuropsychol Dev Cogn B Aging Neuropsychol Cogn. 2017;24(6):662–71.
63. Ryff CD. Happiness is everything, or is it? Explorations on the meaning of psychological well-being. J Pers Soc Psychol. 1989;57:1069–81.
64. Marone S, Bloore K, Sebastiani P, Flynn C, Leonard B, Whitaker K, et al. Purpose in life among centenarian offspring. J Gerontol B Psychol Sci Soc Sci. 2018. https://doi.org/10.1093/geronb/gby023. [Epub ahead of print].
65. Adams ER, Nolan VG, Andersen SL, Perls TT, Terry DF. Centenarian offspring: start healthier and stay healthier. J Am Geriatr Soc. 2008;56(11):2089–92.
66. Kaufman LB, Setiono TK, Doros G, Andersen S, Silliman RA, Friedman PK, et al. An oral health study of centenarians and children of centenarians. J Am Geriatr Soc. 2014;62(6):1168–73.
67. Vasto S, Candore G, Balistreri CR, Caruso M, Colonna-Romano G, Grimaldi MP, et al. Inflammatory networks in ageing, age-related diseases and longevity. Mech Ageing Dev. 2007;128(1):83–91.
68. Galioto A, Dominguez LJ, Pineo A, Ferlisi A, Putignano E, Belvedere M, et al. Cardiovascular risk factors in centenarians. Exp Gerontol. 2008;43(2):106–13.

69. Gubbi S, Schwartz E, Crandall J, Verghese J, Holtzer R, Atzmon G, et al. Effect of exceptional parental longevity and lifestyle factors on prevalence of cardiovascular disease in offspring. Am J Cardiol. 2017;120(12):2170–5.
70. Terry DF, Wyszynski DF, Nolan VG, Atzmon G, Schoenhofen EA, Pennington JY, et al. Serum heat shock protein 70 level as a biomarker of exceptional longevity. Mech Ageing Dev. 2006;127(11):862–8.
71. Olivieri F, Spazzafumo L, Santini G, Lazzarini R, Albertini MC, Rippo MR, et al. Age-related differences in the expression of circulating microRNAs: miR-21 as a new circulating marker of inflammaging. Mech Ageing Dev. 2012;133(11–12):675–85.
72. Walford RL. The immunologic theory of aging. Copenhagen: Munksgaard; 1969.
73. Caruso C, Vasto S. Immunity and aging. In: Ratcliffe MJH, Editor in Chief. Encyclopedia of immunobiology. Vol. 5. Oxford: Academic; 2016. p. 127–32.
74. Campos C, Pera A, Lopez-Fernandez I, Alonso C, Tarazona R, Solana R. Proinflammatory status influences NK cells subsets in the elderly. Immunol Lett. 2014;162:298–302.
75. Potestio M, Pawelec G, Di Lorenzo G, Candore G, D'Anna C, Gervasi F, et al. Age-related changes in the expression of CD95 (APO1/FAS) on blood lymphocytes. Exp Gerontol. 1999;34(5):659–73.
76. Franceschi C, Monti D, Sansoni P, Cossarizza A. The immunology of exceptional individuals: the lesson of centenarians. Immunol Today. 1995;16:12–6.
77. Derhovanessian E, Maier AB, Beck R, Jahn G, Hähnel K, Slagboom PE, et al. Hallmark features of immunosenescence are absent in familial longevity. J Immunol. 2010;185(8):4618–24.
78. Effros RB. Replicative senescence in the immune system: impact of the Hayflick limit on T-cell function in the elderly. Am J Hum Genet. 1998;62:1003–7.
79. Saule P, Trauet J, Dutriez V, Lekeux V, Dessaint JP, Labalette M. Accumulation of memory T cells from childhood to old age: central and effector memory cells in CD4(+) versus effector memory and terminally differentiated memory cells in CD8(+) compartment. Mech Ageing Dev. 2006;127(3):274–81.
80. Ouyang Q, Wagner WM, Voehringer D, Wikby A, Klatt T, Walter S, et al. Age-associated accumulation of CMV-specific CD8+ T cells expressing the inhibitory killer cell lectin-like receptor G1 (KLRG1). Exp Gerontol. 2003;38(8):911–20.
81. Pawelec G, Goldeck D, Derhovanessian E. Inflammation, ageing and chronic disease. Curr Opin Immunol. 2014;29:23–8.
82. Pawelec G. T-cell immunity in the aging human. Haematologica. 2014;99(5):795–7.
83. Pinti M, Appay V, Campisi J, Frasca D, Fülöp T, Sauce D, et al. Aging of the immune system: focus on inflammation and vaccination. Eur J Immunol. 2016;46:2286–301.
84. Pawelec G, Akbar A, Caruso C, Solana R, Grubeck-Loebenstein B, Wikby A. Human immunosenescence: is it infectious? Immunol Rev. 2005;205:257–68.
85. Pawelec G, Akbar A, Caruso C, Effros R, Grubeck-Loebenstein B, Wikby A. Is immunosenescence infectious? Trends Immunol. 2004;25(8):406–10.
86. Pellicanò M, Buffa S, Goldeck D, Bulati M, Martorana A, Caruso C, et al. Evidence for less marked potential signs of T-cell immunosenescence in centenarian offspring than in the general age-matched population. J Gerontol A Biol Sci Med Sci. 2014;69(5):495–504.
87. Buffa S, Bulati M, Pellicanò M, Dunn-Walters DK, Wu YC, Candore G, et al. B cell immunosenescence: different features of naive and memory B cells in elderly. Biogerontology. 2011;12(5):473–83.
88. Bulati M, Buffa S, Candore G, Caruso C, Dunn-Walters DK, Pellicanò M, et al. B cells and immunosenescence: a focus on IgG+IgD-CD27-(DN) B cells in aged humans. Ageing Res Rev. 2011;10(2):274–84.
89. Colonna-Romano G, Bulati M, Aquino A, Pellicanò M, Vitello S, Lio D, et al. A double-negative (IgD-CD27-) B cell population is increased in the peripheral blood of elderly people. Mech Ageing Dev. 2009;130(10):681–90.
90. Colonna-Romano G, Bulati M, Aquino A, Vitello S, Lio D, Candore G, et al. B cell immunosenescence in the elderly and in centenarians. Rejuvenation Res. 2008;11:433–9.

91. Listì F, Candore G, Modica MA, Russo M, Di Lorenzo G, Esposito-Pellitteri M, et al. A study of serum immunoglobulin levels in elderly persons that provides new insights into B cell immunosenescence. Ann N Y Acad Sci. 2006;1089:487–95.
92. Bulati M, Caruso C, Colonna-Romano G. From lymphopoiesis to plasma cells differentiation, the age-related modifications of B cell compartment are influenced by "inflamm-ageing". Ageing Res Rev. 2017;36:125–36.
93. Martorana A, Balistreri CR, Bulati M, Buffa S, Azzarello DM, Camarda C, et al. Double negative (CD19+IgG+IgD-CD27-) B lymphocytes: a new insight from telomerase in healthy elderly, in centenarian offspring and in Alzheimer's disease patients. Immunol Lett. 2014;162:303–9.
94. Rubino G, Bulati M, Aiello A, Aprile S, Gambino CM, Gervasi F, et al. Sicilian centenarian offspring are more resistant to immune ageing. Aging Clin Exp Res. 2019;31(1):125–33.

Individual Longevity Versus Population Longevity

4

Michel Poulain

4.1 Introduction

For scientists, centenarians are examples of healthy ageing [1]. Today, we talk about longevity for a person who is older than 90 years. However, it is often the canonical age of 100 regarded as the threshold of exceptional longevity. Worldwide, the number of centenarians fluctuates between half a million and a million with five women per one man [2]. In the course of the history of humanity, their number has grown [3]; however, centenarians have been observed and confirmed only since the eighteenth century when parish registers and subsequently civil registers helped to validate with certainty the age of the people at the time of their death.

4.2 Individual Longevity

At a global level, the absolute maximum age was reached by Jeanne Calment who died at the grand age of 122 years. Her exceptional age has been often challenged, recently, by a Russian researcher [4]. The record for men is attributed to a Japanese man, Jeroemon Kimura, born in 1897 and deceased on 12 June 2013, at the validated age of 116 years and 54 days [5]. The life expectancy has doubled since the middle of the nineteenth century in most countries, whereas the maximum age at death has progressed more slowly. Geert Adriaan Boomgaard from Groningen died in 1899 at the age of 110 [6] and Margaret Neve from Guernsey at the same age in 1903 [7]. Nowadays, the oldest people on the planet hardly ever reach above 116 years.

M. Poulain (✉)
Université catholique de Louvain, IACCHOS, Louvain-la-Neuve, Belgium

Tallinn University, Estonian Institute for Population Studies, Tallinn, Estonia
e-mail: michel.poulain@uclouvain.be

© Springer Nature Switzerland AG 2019
C. Caruso (ed.), *Centenarians*, https://doi.org/10.1007/978-3-030-20762-5_4

Table 4.1 List of validated oldest olds worldwide based on Gerontological Research Group (www.grg.org) and relayed by Wikipedia (https://en.wikipedia.org/wiki/List_of_the_verified_oldest_people (accessed on January 2019))

	Name	Birth date	Death date	Age	Country
1	Jeanne Calment	21 February 1875	4 August 1997	122 years, 164 days	France
2	Sarah Knauss	24 September 1880	30 December 1999	119 years, 97 days	USA
3	Nabi Tajima	4 August 1900	21 April 2018	117 years, 260 days	Japan
4	Lucy Hannah	16 July 1875	21 March 1993	117 years, 248 days	USA
5	Marie-Louise Meilleur	29 August 1880	16 April 1998	117 years, 230 days	Canada
6	Violet Brown	10 March 1900	15 September 2017	117 years, 189 days	Jamaica
7	Emma Morano	29 November 1899	15 April 2017	117 years, 137 days	Italy
8	Chiyo Miyako	2 May 1901	22 July 2018	117 years, 81 days	Japan
9	Misao Okawa	5 March 1898	1 April 2015	117 years, 27 days	Japan

Lists of the supercentenarians are posted and regularly updated on the Internet by the Gerontological Research Group (www.grg.org), and these lists are relayed by Wikipedia (https://en.wikipedia.org/wiki/List_of_the_verified_oldest_people). The oldest people, deceased or still alive, are ranked by country. Table 4.1 presents the list of the world's oldest elderly, whereas Table 4.2 provides the same list for Italy including those who were still alive in January 2019. The validation of age is an essential prerequisite for any study on centenarians since, without it, the research results are invalid. Demographers insist on a strict validation of age of these exceptional people and contribute within the International Database on Longevity by establishing strict rules [8].

Age is a well-known concept that is chronologically determined by the time-lapse in years and sometimes in days spent between the current date and the date of birth. Even if some parallel concepts have emerged as the biological age determined based on biomarkers or the psychological age estimated based on cognitive aspects, the chronological age is the only exact measurement largely used in research investigations related to ageing. Despite the fact that age is the result of an exact calculation, some errors exist, mostly due to age exaggeration or administrative errors occurring when attributing a specific birth record and birth date to a given person. The validation of age is therefore a crucial task when dealing with scientific study of the determinants of individual longevity.

4.3   The Validation of Age

The validation of the age of alleged centenarians is essential for scientific research in demography, genetics, epidemiology and medicine. Exaggeration of age often observed in past populations is still commonly observed today in populations

Table 4.2 List of validated oldest olds in Italy based on Gerontological Research Group (www.grg.org) and relayed by Wikipedia (https://en.wikipedia.org/wiki/List_of_the_verified_oldest_people (accessed on January 2019))

	Name		Birth date	Death date	Age	Birth region	Death (living) region
1	Emma Morano	F	29 November 1899	15 April 2017	117 years, 137 days	Piedmont	Piedmont
2	Giuseppina Projetto	F	30 May 1902	6 July 2018	116 years, 37 days	Sardinia	Tuscany
3	Maria Giuseppa Robucci	F	20 March 1903	Living	115 years, 221 days	Apulia	Apulia
4	Marie Josephine Gaudette	F	25 March 1902	13 July 2017	115 years, 110 days	USA	Lazio
5	Venere Pizzinato	F	23 November 1896	2 August 2011	114 years, 252 days	Trentino-Alto Adige	Veneto
6	Virginia Dighero	F	24 December 1891	28 December 2005	114 years, 4 days	Liguria	Liguria
7	Maria Redaelli	F	3 April 1899	2 April 2013	113 years, 364 days	Lombardy	Lombardy
8	Maria Gravigi	F	3 March 1900	15 August 2013	113 years, 165 days	Friuli-Venezia Giulia	Friuli-Venezia Giulia
9	Maria Teresa Fumarola	F	2 December 1889	14 May 2003	113 years, 163 days	Apulia	Apulia
10	Lucia Lauria	F	4 March 1896	28 June 2009	113 years, 116 days	Basilicata	Basilicata
11	Stella Nardari	F	23 December 1898	23 February 2012	113 years, 62 days	Veneto	Veneto
12	Ida Frabboni	F	4 October 1896	2 November 2009	113 years, 29 days	Emilia-Romagna	Emilia-Romagna
13	Mariannina Genovese	F	15 October 1905	Living	113 years, 12 days	Sicily	Sicily
14	Diega Cammalleri	F	23 October 1905	Living	113 years, 4 days	Sicily	Sicily
15	Giulia Sani	F	15 September 1893	4 September 2006	112 years, 354 days	Tuscany	Tuscany
16	Antonio Todde	M	22 January 1889	3 January 2002	112 years, 346 days	Sardinia	Sardinia

without reliable civil registration [9]. Several researchers have shown that reporting untrue age is not rare and generally increases with age. This is more often observed in illiterate populations and among males compared with females [10–17].

The basic identification criteria for age validation are those that can be found on the birth record: full name (family name and all given names), sex and date and place of birth. Consequently, the validation of the age of a person, centenarian or not, alive or already dead, consists of checking if a given birth record is accurately attributed to a specific person, so that his or her age may be calculated without doubt. That is ensured by a reliable administrative and civil registration system, but in the case of centenarians, this system would have had to work perfectly for at least 100 years. In some countries, there are some doubts as to whether the birth registration system was working correctly in the nineteenth century. This full registration of persons was first achieved in Sweden in 1749, in Belgium in 1779, in the UK in 1837, in Italy in 1866, in Japan in 1872 and in the USA in 1933. Because it takes more than 100 years for the persons born (or claiming to be born) to be counted, we may consider that the Swedish records achieved accuracy around 1860, at the end of the nineteenth century in Belgium, mid-twentieth century in the UK and more recently in Italy and Japan, while accuracy has yet to be achieved in the USA.

A detailed procedure of age validation is needed. To perform such exercise, we face two different situations depending upon whether the person is still alive or if he or she is already dead. If the person is alive, any appropriate administrative source, such as a population register, a family book, an identification card or a passport, is sufficient to start the age validation process. In no case, a self-declared age can be used as a form of proof for the purposes of age validation. If the person is dead, the death record should provide all the information needed to make a clear identification of the person and ensure the exact linkage with the correct birth record.

In the following paragraphs is presented the summary of the age validation procedure as discussed by Poulain [8].

If the person is dead, the age validation should prove that there is a perfect and unambiguous link between the death and the birth records attributed to this person.

If the person is alive, the age validation should consist of attributing a given birth record to a living person based on all elements of identification, which should be without ambiguity.

The ideal validation procedure will consist of the following steps (see Box 4.1).

Box 4.1

1. *Identification of an alleged centenarian is based on the declaration of age, on a newspaper article, on a special investigation carried out place by place, or on an available official list of inhabitants or centenarians.*
2. *If the alleged centenarian is alive, collect all basic information through identity cards, passports, family or household books, or records and any other pieces of identification that may be available. If the alleged centenarian is dead, locate the death record.*

3. *In both cases, regardless of whether the alleged centenarian is alive or not, locate the birth record.*
4. *Collect all documented life events related to the centenarian, including information on marriage(s), characteristics of the spouse and births of all children.*
5. *Finally, collect all data on births, marriages and deaths of the centenarian's parents and brothers and sisters and identify among the newborns of the period following the birth of centenarian any other newborns with similar names and surnames.*

Many researchers may not be enough rigorous when facing the age validation problem, and even some of them did not consider age validation as necessary when dealing with oldest olds. Researchers should be extremely critical and develop adequate age validation exercises before launching any research project and more specifically when identifying a population with exceptional longevity.

How to conclude a validation process? According to Thoms [18], when the age of a centenarian is being validated, "the proof should be clear, distinct, and beyond dispute". More recently, it has been concluded that it would be more fruitful to discuss the conditions of falsification than to corroborate what may be a false tenet [17]. Like Vincent [19], he does not believe *jusqu'à preuve du contraire*, and therefore supports the strong rules proposed by Thoms [18]. If birth and death certificates exist and are consistent, but only a few pieces of evidence of the history and family reconstruction are collected, it cannot be concluded with "no doubt at all" that the same name in the birth and death records in fact refers to the same person, especially when information is lacking from a large part of the life of the alleged centenarian. All necessary investigations have to be consistent in order to prove validity of age. If one important piece of information is missing, there exists no chance to proceed to validation with "no doubt at all"; but if one key element is wrong, the entire validation process will result negatively. It is definitively easier to prove that this person is not a centenarian than the opposite. In fact, the validation will never be final while the invalidation is generally final when only one clear grounds for invalidation is found. A well-established argument will be sufficient in order to invalidate with high probability the age of an alleged centenarian while a large set of fully consistent documents are needed in order to conclude with high probability that an alleged age has been validated.

4.4 Some Examples of (In)validation of the Age of Alleged Supercentenarians (Extracted from [8])

1. The validation of the alleged age of Antonio Todde of Sardinia. Antonio Todde was declared the oldest documented man on earth in July 2001. Antonio was born in Tiana (Nuoro province, Sardinia) on 22 January 1889, and his birth

record stated that he was the son of Francesco Maria Todde and *della sua unione con donna non maritata* without giving the name of his mother. The missing name for the mother was a negative element for the validation of the exceptional age of Antonio. Fortunately, the baptism record in the parish register brought the missing information and confirmed that Antonio was actually born on 22 January 1889 and was the son of Francesco and Francesca Angela Deiana. In fact, his parents were married according to the church laws but not yet according to the civil registration. Their civil marriage was celebrated only on 30 December 1908 when Antonio was 18 years, and in the marriage record, all brothers and sisters were correctly listed with their age. Antonio died in Tiana on 3 January 2002, a few days before reaching 113 years, and his death record has been correctly linked with his birth record where a marginal note has been added.

2. The invalidation of the alleged age of Damiana Sette of Sardinia. The invalidation of the age of Damiana Sette is instructive for showing that inadequate documentation allowed her to be considered a supercentenarian when, in fact, she was not. Maria Angelica Damiana Sette died in Villagrande at the alleged age of 110 on 25 February 1985. According to her death record, she was born in Villagrande on 8 August 1874, the daughter of Pietro Sette and Monserrata Pirroni. A birth record was found for a child named Maria Angelica Damiana Sette born in Villagrande on 8 August 1874, daughter of Pietro Sette and Monserrata Pirroni. The death record is wholly compatible with this birth record. In the birth record, a marginal note is found of the 25 February 1985 death linking by the civil registration officer both events and records. Consequently, Damiana Sette was considered a true supercentenarian.

   At this stage of the validation process, everything would seem to confirm the fact that Damiana Sette did die at the venerable age of 110, information transcribed on her gravestone in the Cemetery of Villagrande. However, the meticulous reconstruction of the family composition of Damiana Sette, based on the civil status registers and the *anagrafe*, allows one to conclude that the person who died in 1985 was not Maria Angelica Damiana Sette born 8 August 1874 but her younger sister called Maria Monserrata Damiana Sette who was born on 5 May 1877. The real death record of Maria Angelica Damiana Sette, in this document named Angelica, has been found and was dated 10 June 1876 at the age of 22 months. This type of error occurs frequently in historical demography when reconstructing family profiles that link different data related to births and deaths. This is due to the fact that when a child died at a young age, it was customary in some cultures to consider that the child with the same sex born immediately after this death in some way will replace the deceased. Therefore, the next child was given the same forename or at the very least certain identical forenames. This was the case for Maria Angelica Damiana Sette who died at the age of 22 months and was "replaced" by her younger sister called Maria Monserrata Damiana Sette. This administrative error remained until her death, and consequently, the marginal annotation relating to her death appeared on the death certificate of her older sister, and given names were those of her older sister, although she was usually called Damiana.

3. The validation of alleged age of Johan Riudavets of Menorca. The age of Johan Riudavets, the oldest man on earth after the death of Yukichi Chuganji on 28 September 2003, was easy to validate because administrative pieces fit perfectly

with the declared age. Information obtained through direct interviews with
Johan and his daughter, Francesca, as well as additional documentation showing Johan as head of local youth in 1912 when he was 23 years, are compatible,
proving the accuracy of the birth record. As observed, Johan's family is experiencing exceptional longevity, a fact that reinforces the accuracy of the age of
Johan Riudavets who died on 5 March 2004, some months after he celebrated
his 114th birthday.

4. The Case of Kamato Hongo of Japan. The example of Kamato Hongo is instruc-
   tive regarding age validation. Carefully analysed documents related to Kamato
   Hongo revealed an important error. Kamato, reported to be the second daughter
   and fifth child in the family, is transcribed in the fifth position on the *Koseki
   Shohon*, i.e., the register of family members, births, deaths, adoptions and so on.
   It is essentially a birth, death and marriage certificate combined into one form.
   According to her official date of birth, she should be in fourth position, and her
   older sister was born only 7 months before her. At that stage of the validation,
   strong arguments favour invalidation of her age, but there was one possibility for
   being validated, for example, if Kamato was an adopted child who arrived in the
   family in fifth position when she was already aged more than 3 years. But the
   Koseki does not mention this hypothetical adoption, and moreover, when comparing these two life histories of Kamato Hongo, it would sound better if she was
   the last child born around 1893, as the average interval between successive births
   is about 3 years and even more if this is the last child.

The Kamato Hongo life history appears as following according to available
documents:

> *Kamato was born in 1887 when her mother was supposed to be aged 43. She had two chil-
> dren before marriage when she was 22 and 25 years. She married in 1914 when she was
> 27 years, even though she told others that she was not very old when she married. Then she
> had her five other children from age 29 up to 45, and finally, she died at 116 in October
> 2003.*

However, the following life history definitively sounds more plausible, particularly when we consider that during her interview she and her daughter confirmed
that Kamato did not marry as late as age 27.

> *Kamato was born in 1893 when her mother was supposed to be aged 49. She had two chil-
> dren before marriage when she was 16 and 19 years. She married in 1914 when she was
> 21 years, and the two first children have been recognized at the same time. Then she had her
> other children up to age 39, and finally she died at 110 years in October 2003.*

4.5   About the Age of the Oldest Person on Earth

Since the beginning of the twentieth century when Margaret Neve took the record
of oldest validated person from the hands of Geert Boomgaard [6, 7], the oldest
person on earth has always been a woman as shown on the list of record holder
presented in Table 4.3 based on the GRG database (http://www.grg.org).

Table 4.3 List of validated oldest persons on earth since the dawn of the twentieth century (http://www.grg.org, accessed on January 2019)

Name	Years	Days	Born	Died
Geert Adriaans Boomgaard	110	135	21 September 1788	3 February 1899
Margaret Ann (Harvey) Neve	110	321	18 May 1792	4 April 1903
Delina (Ecker) Filkins	113	214	4 May 1815	4 December 1928
Fannie Leona Thomas	113	273	24 April 1867	22 January 1981
Augusta Louise (Hoppe) Holtz	115	79	3 August 1871	21 October 1986
Jeanne Calment	122	164	21 February 1875	4 August 1997

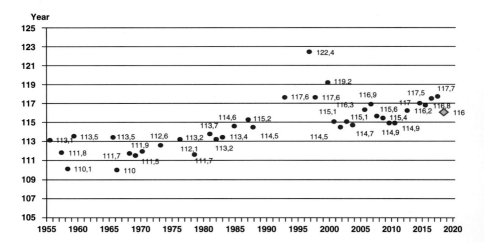

Fig. 4.1 Evolution of the oldest woman on earth by birth cohort (1865–1906) (http://www.grg.org)

The evolution of the highest age achieved by birth cohort is presented in Fig. 4.1. The trend shows a relative *statu quo* for the generation born before 1870; thereafter, the increase is evident until 1890 even if we exclude Jeanne Calment. Further, the improvement is largely reduced. In fact, more people reach the exceptional age of 115 years, but the highest age achieved does not increase as well. This observation is confirmed by the relative stability of average age at death of the ten oldest women for each birth cohort born between 1894 and 1903 (Table 4.4).

The role of the size of the population concerned is also important to explain the variation of the oldest age achieved between the different countries. As shown in Fig. 4.2, for women only, a higher age achieved is observed if the population of the country is large. A ten-time larger population means, statistically spoken, three more years for the maximum age achieved by the population in a given country.

Table 4.4 Average age for the ten oldest women worldwide by birth cohort (http://www.grg.org, accessed on January 2019)

Cohorts	Average
1894	113.1
1895	114.0
1896	114.6
1897	113.9
1898	114.1
1899	113.8
1900	115.2
1901	114.7
1902	114.1
1903	114.7

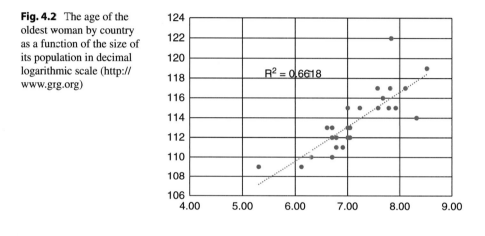

Fig. 4.2 The age of the oldest woman by country as a function of the size of its population in decimal logarithmic scale (http://www.grg.org)

4.6   Population Longevity

For a long time, longevity was studied at the individual level separately considering the phenomenon of longevity of each centenarian. Once the age validation was positively performed, researchers focused mainly their efforts on finding the determinants of this individual longevity. The best known studies are as follows: in the USA, the Georgia Centenarian Study [20] and the New England Centenarian Survey (http://www.bumc.bu.edu/centenarian/); in Northern Europe, the study on centenarians in Sweden [21] and in Denmark [22]; and in Japan, the Okinawa Centenarian Study (www.okicent.org) and the Tokyo Centenarian Study (http://cheba.unsw.edu.au/content/tokyo-centenarian-study). In France, it is worth to mention the IPSEN survey that established the characterization of 1000 French centenarians [23]. All these research groups collaborate within the International Centenarian Consortium and contribute to identifying specific features among centenarians worldwide.

Population longevity is related to a population, whereas individual longevity concerns individuals. As for individual longevity, the validation of age is the first step when identifying a population experiencing higher longevity. A low life expectancy, a high prevalence of centenarians, the existence of persons alive or dead with an exceptional extreme age, a high proportion of illiterates and an unusual age and sex structure among centenarians are contextual factors that suggest a low probability that the stated age of a centenarian is accurate in a given population. This is often the case for the remote populations in mountainous regions and on small islands. Yet, some populations have been identified where longevity is shared by a large portion of their members. When in a given population the proportion of people surviving to oldest ages is greater than in neighbouring areas, according to a predetermined threshold deviation, it is referred as population longevity.

The reasons why the interest towards the study of long-living populations has been so far limited are the following ones: (1) The identification of a long-living population and the age validation of its exceptional members are time-consuming, and the requested tasks are not easy to fulfil mostly due to the lack of documentary evidence. (2) Communities experiencing higher population longevity are often small-sized and the oldest old group may consist of a limited number of cases increasing the risk of low statistical power in analysis. (3) More generally, the search for population longevity determinants requires the mastering of several disciplines, a relatively complicated task, as specific methods have still to be developed. A consensus strategy of analysis consistent with data collected in various disciplines has not been reached yet.

When searching for the determinants of population longevity, the relevant characteristics or behaviours are those shared by a large part of the population. By considering these common characteristics, the chance to find more powerful explanatory variables is increased as most persons concerned are born and live in the same place. Therefore, they are more likely to share the same genetic makeup, early life conditions, as well as traditional behaviours and habits, including the same locally produced food. Therefore, by identifying areas where people live longer, the search for longevity determinants could be improved.

4.7   The Blue Zones

Conventionally, the apparent exceptional longevity of a population is inferred from the existence of an unusually large proportion of centenarians/nonagenarians. To compare longevity levels, the centenarian prevalence (CP), i.e., the ratio between the number of living centenarians in a given population and the total resident population, is largely used by the scientists as well as by the media. However, the reliability of this indicator deserves critical evaluation, as it is sensitive to a number of biases related to existing migration flows and changes in fertility behaviour. For example, in case of a population that experienced either a baby boom or a large-scale immigration flow of young people, the CP will fail to identify remarkable survival of persons in old ages, as the proportion of elderly is artificially lowered. On the contrary, where the younger population has dropped due to emigration, the proportion of elderly may result artificially increased. In such cases, the prevalence or proportion of oldest olds is no longer reliable to measure longevity and should not be used for comparison across populations. Nevertheless, CP is still the most frequently used indicator by gerontologists as well as by national and regional authorities, eager to claim a longevity status for the area of the concerned population. Population longevity can only be identified through various indexes that are mostly associated with the cohort life tables computed for that population as a whole. To compare extreme population longevity between two populations, the best indicator is the probability to reach 100 years among the babies born in that given population a century before. Nevertheless, the so-called Extreme Longevity Index [24] is hard to compute because of the lack of requested data and the disturbing effect of migration flows.

Figure 4.3 displays the prevalence of centenarians by region, in Italy, at the 2011 census. Northern Italy, in particular Liguria, has the highest prevalence of centenarians. This interpretation is biased. Firstly, alongside the last century, the fertility of women was systematically lower in Northern Italy compared with Southern Italy including both Sardinia and Sicily. Secondly, important differences in the migration flows related to young or old people largely affected the reliability of the prevalence. Thus, these figures are usually misleading to compare the variation of the level of longevity between Italian regions.

If the number of centenarians is reported to the number of newborns in the same region a century ago, the disturbing effect of variation in fertility is controlled, but still the important differences in migration flows are not considered. In Fig. 4.4, Lazio, the most attractive region for migrants, shows a largely higher prevalence, whereas the regions of Southern Italy that experienced strong emigration flow, especially Sicily and Calabria, show an underestimated level of longevity. These two comparative exercises aim to demonstrate the difficulties faced when comparing the level of longevity between different populations.

Based on the proportion of centenarians enumerated in censuses or mortality rates, the first potential longevity areas were identified at the beginning of the

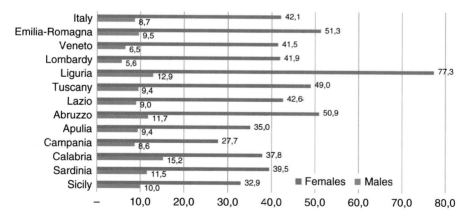

Fig. 4.3 Prevalence of centenarians by region of residence in Italy at 2011 census (Available from https://www.istat.it/it/censimenti-permanenti/censimenti-precedenti/popolazione-e-abitazioni/popolazione-2011 (in Italian))

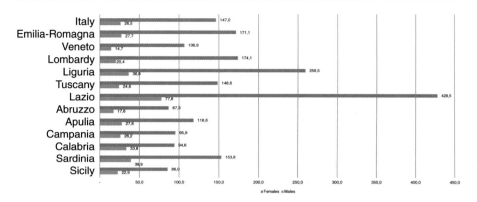

Fig. 4.4 Prevalence of centenarians by region of birth in Italy at 2011 census (Available from https://www.istat.it/it/censimenti-permanenti/censimenti-precedenti/popolazione-e-abitazioni/popolazione-2011 (in Italian))

twentieth century. The US Census Bureau compared the proportion of centenarians in different countries and pointed out the exceptional cases of Bolivia with 75 centenarians per 100,000 in 1900, Bulgaria with 60 in 1905 and the Philippines with 51 in 1903 [10].

In the January 1973 issue of National Geographic magazine, the physician Alexander Leaf gave a detailed account of his journeys to countries of purported long-living people: the Hunzas from Pakistan, the Abkhazians from the Soviet Union and the Ecuadorians from Vilcabamba. According to Leaf, there were ten times more centenarians in these countries than in most Western ones. He pointed out that each of these countries was characterised by "poor sanitation, infectious diseases, high infant mortality, illiteracy and a lack of modern medical care making the inhabitants extreme longevity even more extraordinary" [25]. Some years later, Mazess and Forman showed that age exaggeration was predominant in Vilcabamba with a large number of the men and women who tended to increase their age to improve their social status or to promote local tourism [12]. Later, Leaf acknowledged this conclusion and made a final statement agreeing that no evidence proved the unusual ages in the village of Vilcabamba [26]. Moreover, for both Hunzas from Pakistan and Abkhazians in Caucasus, there was no documentation validating the exceptional alleged age of centenarians.

Since the end of the previous century, demographers have become increasingly concerned with the accuracy of longevity claims, given the unprecedented rise in very old people in developed countries [9, 17]. An international group of researchers developed the International Database on Longevity, aiming to validate cases of alleged exceptional longevity around the world. Age misreporting and, more specifically, age exaggeration must be ruled out [27]. Consequently, more careful checks of all documents, interviews and available statistical data have been conducted. This has resulted in a systematic invalidation of all allegedly long-living populations worldwide, as most claims of extreme age for their inhabitants appeared to be undocumented or exaggerated. Perls et al. explained, "there are several geographical areas that have claimed inhabitants with extreme longevity", and would be therefore considered as longevity areas, "but after closer examination these claims have been found to be false" [28]. He concluded, "such cases of extreme longevity required detailed scrutiny because they are so incredibly rare". Young et al. [9] listed various cases of invalidation of extreme ages and false longevity hot spots.

In addition to the validation rules for individual centenarians, the validation of an extreme population longevity requires several specific investigation steps that vary depending on the current availability of data sources for each specific population. To assess the level of population longevity that is estimated through the average individual longevity within a given population, data on all births and deaths occurring within this population must be collected. The main objectives are to ensure a comprehensive collection of birth data a century ago as well as data on currently surviving centenarians and data on deaths of centenarians during the last decades.

In 1999, following the publication of data showing extreme male longevity in Sardinia [29], the scepticism pushed demographers to assess the validity of the alleged ages of the oldest olds in Sardinia [30]. Based on a strict validation method, the ages of Sardinian centenarians were thoroughly checked and proved correct [27]. This validation was based on investigations in the civil registers of births from the last two decades of the nineteenth century and the registers of deaths from the last two decades of the twentieth century. Considering the marginal annotation on death found in the birth registers, all centenarians were identified by place of birth.

Surprisingly, the spatial distribution of Sardinian centenarians according to their place of birth was far from random. In March 2000, in the course of this validation process, an area was identified in the mountains where the proportion of centenarians calculated as previously stated was significantly higher than in the surrounding areas (Fig. 4.5a, b) [24]. This area was called the Longevity Blue Zone or shortly Blue Zone (BZ). The term BZ was chosen because a blue pen was used in March 2000 to surround the villages with long-living population on the map of Sardinia. Since then, an area where the population characterized by a significantly higher level of longevity compared with neighbouring regions is defined as being a BZ, provided that the exceptional longevity of people in this population has been fully validated. In practice, a BZ is conceptually a rather limited and homogenous geographical area where the population shares the same lifestyle and environment and its longevity has been proved to be exceptionally high [31].

To identify a BZ and to prove the exceptional longevity of its population, it is necessary first to validate the individual longevity of people living in the candidate area and more precisely to assess accurately the age at death or the extreme survival of oldest old. Thereafter, the number of centenarians born in the area was reported to the number of newborns in the same area a century earlier. The spatial distribution of centenarians by place of birth smoothed by a Gaussian method appeared to be fully different from the original one that considered the distribution of centenarians by place of residence without considering the variation in the size of the population. The validation process allowed identifying an area in the mountainous part of Sardinia with a significantly higher proportion of centenarians out of population born in the same place.

The extreme longevity area identified in the mountainous part of Sardinia includes a group of 14 villages in the Barbagia and Ogliastra districts, encompassing a population of about 40,000 inhabitants, mainly engaged in pastoral and agricultural activities, and following a relatively traditional lifestyle. This population remained isolated for centuries, which contributed to make its gene pool more homogeneous and to the preservation of sociocultural and anthropological characteristics throughout its history [32].

In 2004, Dan Buettner, an American journalist, travelled to Sardinia and to Okinawa, another BZ [31], and wrote a paper published by National Geographic in

Fig. 4.5 (a) The spatial distribution of surveyed centenarians in the AKEA project [29]. From [31] with permission. (b) The first Blue Zone as identified thereafter [24]. From [31] with permission

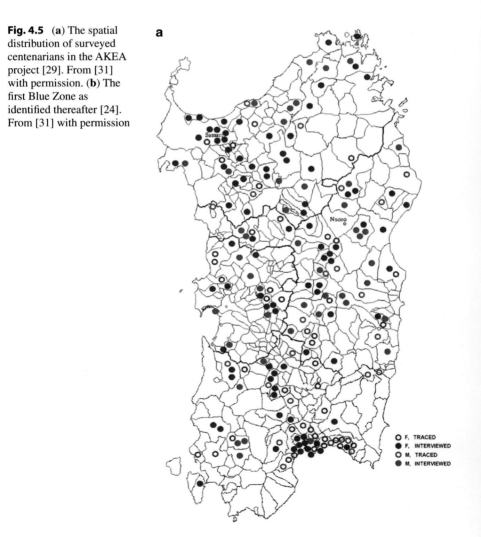

b

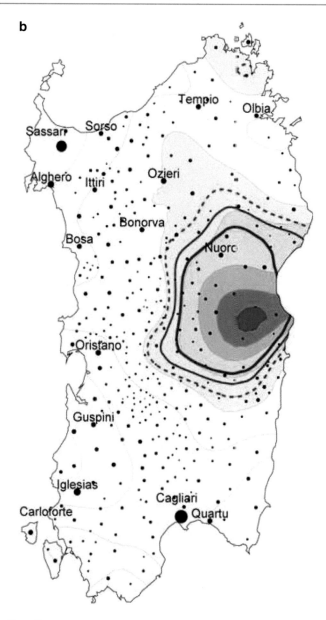

Fig. 4.5 (continued)

November 2005 [33] on these two longevity hot spots. He added the Adventist community of Loma Linda as example of long-living population on the territory of the USA. Hence, this community cannot be considered as being a BZ sensu stricto alongside the concept presented above. Later, following the work initiated by Luis Rosero-Bixby [34] in Costa Rica, the Nicoya peninsula was identified as corresponding to the criteria of BZ and the same was found for the island of Ikaria in Greece. On the basis of these scientific investigations, Buettner largely disseminated the lessons of the BZ [35] and developed public initiatives to transfer these lessons to help people to live longer and better in several US communities (see www.bluezonesproject.com).

Currently, the populations of each of these four BZ are studied by several groups of researchers aiming to trace the determinants of this phenomenon [31]. When seeking population longevity determinants, the relevant characteristics or behaviours are those shared by a large part of the population. The chance of finding more powerful explanatory variables is increased because many people were born and lived in the same place. Thus, they are more likely to share genetic makeup. Considering possible explanations, BZ populations live in islands and/or mountainous regions; thus, they are geographically and/or historically isolated and live in an unpolluted environment. These populations have managed to maintain a traditional lifestyle. This implies reduced level of stress, an intense family and community support for the elderly and the consumption of locally produced food. They are also characterized by an intense physical activity that goes beyond 80 years; this is in agreement with evidence from studies that support the positive association between increased levels of physical activity, exercise participation and improved health in older adults [36]. This has probably enhanced the accumulation of factors limiting the conditions that have a negative impact on health in Western world. In these populations, the remarkably good state of health during ageing could be the result of a delicate balance between the benefits of modernity, as the increase in wealth and better medical cares, and those of a traditional lifestyle. All these factors could have promoted an ideal environment for the emergence of long-lived phenotypes at the population level. By studying the various BZ populations, therefore, identification of the causal factors of longevity might be enhanced.

4.8   Conclusion

The study of the longevity of the specific populations living in the so-called Blue Zones aims at seeking determinants shared by a sizable percentage of people within a given population. The determinants of population longevity could be either individual characteristics (factors or traits that are associated with individual longevity) or contextual factors related to the global environment shared by that population and responsible for population longevity. A conventional study centred on individual longevity is hardly able to capture these "shared" components peculiar of population longevity. The analysis of population longevity should not be viewed as an alternative to individual longevity, but it could potentially complement the more conventional approaches and facilitate the identification of longevity factors acting at the super-individual level.

References

1. Engberg H, Oksuzyan A, Jeune B, Vaupel JW, Christensen K. Centenarians—a useful model for healthy aging? A 29-year follow-up of hospitalizations among 40,000 Danes born in 1905. Aging Cell. 2009;8:270–6.
2. Herm A, Cheung SLK, Poulain M. Emergence of oldest old and centenarians: demographic analysis. Asian J Gerontol Geriatr. 2012;7:19–25.
3. Robine J-M, Caselli G. An unprecedented increase in the number of centenarians. Genus. 2005;61:57–82.
4. Zak N. Evidence That Jeanne Calment Died in 1934. Rejuvenation Res. 2019;22(1):3–12. https://doi.org/10.1089/rej.2018.2167.
5. Gondo Y, Hirose N, Yasumoto S, Arai Y, Saito Y. Age verification of the longest lived man in the world. Exp Gerontol. 2017;99:7–17.
6. Chambre D, Jeune B, Poulain M. Geert Adriaan Boomgaard, the first supercentenarian in history? In: Maier H, Jeune B, Vaupel JW, editors. Exceptional lifespans; 2019.
7. Poulain M. Margaret Ann Harvey Neve—110 years in 1903. The first documented female supercentenarian. In: Maier H, Jeune B, Vaupel JW, editors. Exceptional lifespans; 2019.
8. Poulain M. On the age validation of supercentenarians. In: Maier H, Gampe J, Jeune B, Robine J-M, Vaupel JW, editors. Supercentenarians. Heidelberg: Springer; 2010. p. 3–30.
9. Young RD, Desjardins B, McLaughlin K, Poulain M, Perls TT. Typologies of extreme longevity myths. Curr Gerontol Geriatr Res. 2010;2010:423087.
10. Bowerman WG. Centenarians. Transactions of the Actuaries Society of America. 1939;40:361–78.
11. Myers RJ. Validation of centenarian data in the 1960 census. Demography. 1966;3:470–6.
12. Mazess RB, Forman SH. Longevity and age exaggeration in Vilcabamba, Ecuador. J Gerontol. 1979;34(1):94–8.
13. Rosenwaike I. A new evaluation of United States census data on the extreme aged. Demography. 1979;16:279–88.
14. Palmore EB. Longevity in Abkhazia: a reevaluation. Gerontologist. 1984;24:95–6.
15. Bennett NG, Garson LK. Extraordinary longevity in the Soviet Union: fact or artifact? Gerontologist. 1986;26:358–61.
16. Jeune B. In search of the first centenarians. In: Jeune B, Vaupel JW, editors. Exceptional longevity: from prehistory to the present, Odense monographs on population aging, vol. 2. Odense: Odense University Press; 1995. p. 11–24.
17. Jeune B, Vaupel JW, editors. Validation of exceptional longevity, Odense monographs on population aging, vol. 6. Odense: Odense University Press; 1999. p. 249.
18. Thoms WJ. Human longevity, its facts and its fictions. London: John Murray; 1873.
19. Vincent P. La mortalité des vieillards. Population. 1951;6:181–204.
20. Poon LW, Clayton GM, Martin P, Johnson MA, Courtenay BC, Sweaney AL, et al. The Georgia Centenarian Study. Int J Aging Hum Dev. 1992;34:1–17.
21. Samuelsson SM, Alfredson BB, Hagberg B, Samuelsson G, Nordbeck B, Brun A, et al. The Swedish Centenarian Study: a multidisciplinary study of five consecutive cohorts at the age of 100. Int J Aging Hum Dev. 1997;45:223–53.
22. Andersen-Ranberg K, Schroll M, Jeune B. Healthy centenarians do not exist, but autonomous centenarians do: a population-based study of morbidity among Danish centenarians. J Am Geriatr Soc. 2001;49:900–8.
23. Allard M, Robine J-M. Les centenaires français. Etude de la fondation IPSEN. Paris: Serdi Editores; 2000.
24. Poulain M, Pes GM, Grasland C, Carru C, Ferrucci L, Bagg o G, et al. Identification of a geographic area characterized by extreme longevity in the Sardinia island: the AKEA study. Exp Gerontol. 2004;39:1423–9.

25. Leaf A. Every day is a gift when you are over 100. Natl Geogr. 1973;143:93–118.
26. Leaf A. Statement regarding the purported longevous peoples of Vilcabamba. In: Hershow H, editor. Controversial issues in gerontology. New York: Springer; 1981. p. 25–6.
27. Poulain M, Deiana L, Ferruci L, Pes GM, Carru C, Franceschi C, et al. Evidence of an exceptional longevity for the mountainous population of Sardinia. In: Robine J-M, Horiuchi S, editors. Human longevity, individual life duration, and the growth of the oldest-old population. New York: Springer; 2006.
28. Perls T, Kunkel LM, Puca AA. The genetics of exceptional human longevity. J Am Geriatr Soc. 2002;50:359–68.
29. Deiana L, Ferrucci L, Pes GM, Carru C, Delitala G, Ganau A, et al. AKEntAnnos. The Sardinia study of extreme longevity. Aging (Milano). 1999;11:142–9.
30. Koenig R. Demography. Sardinia's mysterious male Methuselahs. Science. 2001;291:2074–6.
31. Poulain M, Herm A, Pes GM. The blue zones: areas of exceptional longevity around the world. Vienna Yearb of Popul Res. 2013;11:83–102.
32. Pes GM, Tolu F, Dore MP, Sechi GP, Errigo A, Canelada A, et al. Male longevity in Sardinia, a review of historical sources supporting a causal link with dietary factors. Eur J Clin Nutr. 2015;69:411–8.
33. Buettner D. The secrets of longevity. Natl Geogr. 2005:5–26.
34. Rosero-Bixby L. The exceptionally high life expectancy of Costa Rican nonagenarians. Demography. 2008;45:673–91.
35. Buettner D. The blue zones: 9 lessons for living longer from the people who've lived the longest. Washington, D.C.: National Geographic Books; 2008.
36. Pes GM, Dore MP, Errigo A, Poulain M. Analysis of physical activity among free-living nonagenarians from a Sardinian Longevous population. J Aging Phys Act. 2018;26:254–8.

Dietary Inflammatory Index in Ageing and Longevity

5

Luca Falzone, Massimo Libra, and Jerry Polesel

5.1 Introduction

Over the last century, several scientific findings in the epidemiological, biochemical, biological and molecular fields have allowed to identify numerous risk factors responsible for the development of various human diseases including metabolic disorders, cardiovascular diseases (CVDs), cancer and chronic degenerative diseases. Among these risk factors, a fundamental role is played by ageing, inflammation and diet, three interrelated risk factors able to modify the homeostasis of the organism leading to the onset of several age-related diseases [1, 2].

Ageing is a natural physiological process due to the loss of cell functions and tissue renewal caused by cell senescence and reduction in the number of stem cells and tissue plasticity, respectively [3, 4]. In the last decades, the increase in life expectancy has shown that ageing has led to an increase in the incidence rates of many diseases. Behind the increase in the incidence of age-related disorders, there is the loss of control mechanisms of different cellular processes including proliferation, apoptosis and senescence. In particular, ageing is characterized by the accumulation

L. Falzone
Department of Biomedical and Biotechnological Sciences, University of Catania,
Catania, Italy
e-mail: luca.falzone@unict.it

M. Libra (✉)
Department of Biomedical and Biotechnological Sciences, University of Catania,
Catania, Italy

Research Center for Prevention, Diagnosis and Treatment of Cancer, University of Catania,
Catania, Italy
e-mail: mlibra@unict.it

J. Polesel
Unit of Cancer Epidemiology, Centro di Riferimento Oncologico di Aviano (CRO) IRCCS,
Aviano, Italy
e-mail: polesel@cro.it

© Springer Nature Switzerland AG 2019
C. Caruso (ed.), *Centenarians*, https://doi.org/10.1007/978-3-030-20762-5_5

of genetic mutations mostly due to loss of DNA repair mechanisms, by the loss of telomeric sequences, by the inappropriate turnover of proteins, by epigenetic alterations (e.g. altered methylation patterns and de-regulation of microRNAs) and by mitochondrial dysfunction [5]. In addition, ageing is associated with alterations of inflammatory processes and oxidative stress that determine the increase in molecular changes [6]. In particular, the close relationship between inflammation and ageing has led the scientific community to define the inflammatory status present in the elderly with the term "inflamm-ageing" [7]. Inflamm-ageing is responsible for a constant low-grade chronic and systemic inflammation that deteriorates both cells and tissues, leading to the development of age-related pathologies including cancer, CVDs and type 2 diabetes (T2DM). All these pathologies share a common inflammatory background characterized by high levels of pro-inflammatory cytokines and mediators, such as interleukin (IL)-6 and C-reactive protein (CRP) [6].

With ageing, the metabolism of nutrients introduced with diet changes drastically. In particular, in recent decades, especially in the most developed countries, there has been a drastic increase in the number of overweight and obese people. The major increase was mainly observed in the sixth to seventh decades of life, as ageing predisposes to the accumulation of visceral fat and consequently to the increase in weight [8]. Through biochemical studies, it has been possible to establish that obesity is responsible for a mild but chronic inflammation, which represents a constant stress for the body [9].

An excess of macronutrients introduced through the diet determines an increased accumulation of body and circulating fatty acids harmful for the organism [10]. Consequently, obesity and the excess of circulating fatty acids lead to the stimulation of both macrophages and adipocytes with the triggering of different signal transduction pathways involved in the pro-inflammatory status [11]. Specifically, in obese subjects, there is an evident increase in pro-inflammatory cytokines and a reduction in adiponectin levels that lead to an increase in the production of reactive oxygen species (ROS) and a decrease in endothelial nitric oxide [11]. This chronic inflammatory status is called metaflammation, defined as "low-grade, chronic inflammation orchestrated by metabolic cells in response to excess nutrients and energy" [12, 13]. Metaflammation is not only the result of overweight and obesity, but it also depends on ageing [6]. Both metaflammation and inflamm-ageing are responsible for a bivalent cycle where the pro-inflammatory status induced by metaflammation and inflamm-ageing leads to cellular damage and senescence. On the other hand, senescent cells produce several growth factors, proteases, chemokines and pro-inflammatory cytokines that worsen the chronic inflammatory status, creating a vicious circle that, in the long period, determines the occurrence of different pathological conditions including age-related diseases [14]. In fact, the molecular pathways altered in both diet- and obesity-induced metaflammation and inflamm-ageing are signal transduction pathways whose alterations are associated with an increased risk of development of age-related diseases [15, 16].

In addition, both ageing and inflammation are sensitive to several environmental factors, among which diet represents the most powerful regulating determinant [17]. In particular, the diet, especially the individual foods that compose the diet, is able to change the body's response to pro-inflammatory stimuli and to limit or change the ageing process [18].

For this purpose, in order to quantify the impact of specific food regimes on the inflammatory profile of the individual, a meter of the pro-inflammatory and anti-inflammatory power of foods, called "Dietary Inflammatory Index" (DII), has been developed [19].

Foods, through their inflammatory power, are able to modify cellular homeostasis and consequently different cellular processes including proliferation, apoptosis and senescence. There are numerous studies demonstrating how diet can play a fundamental role in ageing and anti-ageing, counteracting the formation of a pro-inflammatory cellular environment [20]. Indeed, whilst foods with a high inflammatory index lead to senescence and oxidative stress harmful to cells, it is true that foods with a low inflammatory index and functional foods can reduce oxidative stress thanks to antioxidants introduced with the diet, i.e. nutraceuticals (Chap. 11), which are very effective in counteracting both ageing and the onset of age-related diseases [21, 22].

More in detail, a diet characterized by foods with a high DII is responsible for the production of several cytokines able to negatively stimulate epithelial cells for the production of ROS and to induce the activation of the signalling pathway responsible for the decision of the cell fate [23, 24]. Moreover, a diet with a high DII triggers the production of multiple secreted inflammatory cytokines, their cognate receptors and positive-feedback loops with corresponding transcription factors as key mediators of cell senescence [25]. All these molecular alterations result in the accumulation of DNA damages that, in turn, induce the activation of the NF-κB signalling pathway, resulting in the cell cycle arrest and the induction of a senescent phenotype [25, 26].

In the following paragraphs, it is emphasized how the DII has allowed to accurately establish the relative risk of developing cardiovascular, metabolic and oncological diseases, taking into consideration not only the inflammatory power of food but also the pathogenetic role of ageing in favouring the development of these pathologies. It is also be analysed how dietary habits affect ageing and longevity processes of individuals and how the DII can be used as a tool to measure the anti-ageing potential of diet.

5.2   The Dietary Inflammatory Index

The DII is a literature-derived tool developed to have a comprehensive evaluation of the inflammatory potential of diet [19]. The DII has been validated in a variety of longitudinal and cross-sectional studies using various inflammatory markers including CRP, IL-6 and tumour necrosis factor-α [19, 27, 28].

Briefly, DII is based on a literature review of the association between dietary habits and inflammatory biomarkers, which identified and scored 45 foods or nutrients derived from usual diets [19, 29]. An inflammatory effect score was available for each food parameter, ranked according to inflammatory potential in anti-inflammatory (negative scores) and pro-inflammatory (positive scores) food parameters. Table 5.1 reports some examples of anti- and pro-inflammatory foods and nutrients included in the DII calculation, as originally reported by Shivappa and colleagues [19]. A person's usual intake of each food parameter is weighted according to its inflammatory effect score and then summed to produce the overall DII. DII score increases with increasing inflammatory potential of diet, with negative values indicating anti-inflammatory diets and positive ones indicating pro-inflammatory diets.

Although originally developed to identify the inflammatory potential of diet [9, 29], the DII has the additional property to indicate a general nutritional quality of diet.

Table 5.1 Food and nutrients included in the calculation of the Dietary Inflammatory Index (DII) according to their inflammatory potential

Food parameter	Unit of measure	Inflammatory effect score
Anti-inflammatory food parameters		
Garlic	G	−0.412
Onion	G	−0.301
Vitamin B_6	Mg	−0.365
Polyunsaturated fatty acids	G	−0.337
Fibre	G	−0.663
β-Carotene	μg	−0.584
Folic acid	μg	−0.190
Magnesium	Mg	−0.484
Niacin	Mg	−0.246
Vitamin A	RE	−0.401
Vitamin C	Mg	−0.424
Vitamin D	μg	−0.446
Vitamin E	Mg	−0.419
Pro-inflammatory food parameters		
Energy	Kcal	0.180
Carbohydrate	G	0.097
Protein	G	0.021
Total fat	G	0.298
Saturated fat	G	0.373
Cholesterol	Mg	0.110
Iron	Mg	0.032
Vitamin B_{12}	μg	0.106

Indeed, several studies reported a positive association between anti-inflammatory diets and adequate intake of nutrients. Results from the Whitehall II cohort study, a cohort study of men and women originally employed by the British civil service in London-based offices, showed that people taking anti-inflammatory diets reported higher intake of fibres, proteins and polyunsaturated fatty acids and lower intake of carbohydrates, fats and fatty acids than people with pro-inflammatory diets [30]. Similarly, in the ORISCAV-LUX cohort, a cross-sectional study on the prevalence of cardiovascular risk factors among the adult population of Luxembourg, aged 18–69 years, people taking anti-inflammatory diets reported higher intake of vitamins and minerals than those taking pro-inflammatory diets [31]. Further, an association between DII and other indicators of healthy diets has been reported [30, 32]. In particular, anti-inflammatory scores have been consistently reported in people who follow a Mediterranean Diet, whose beneficial effects on healthy ageing and longevity are discussed in Chap. 10 [33–35].

Bearing in mind this second DII property, it is of great interest to understand how DII varies across age groups, with a particular focus on the elderly. Figure 5.1 shows DII scores for Italian men and women enrolled in population-based studies [35].

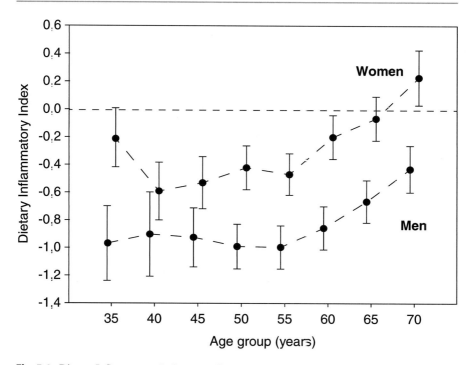

Fig. 5.1 Dietary Inflammatory Index according to gender and age in Italian cancer-free people

Both genders reported anti-inflammatory diets (i.e. negative DII scores), even if DII score in men were consistently lower than that in women. The score starts increasing from the age of 55 to 59 years, with a similar trend in both men and women. Although some differences emerged across geographic areas, the rising trend with age in the elderly was consistent [35]. The increase in the inflammatory potential of diet with age could have relevant health implication: in addition to its impact on several clinical outcomes, it could indicate a general worsening of diet in the elderly. An adequate nutritional status is important to prevent frailty in this vulnerable population, thereby the DII score could be a useful tool to identify people with inadequate diet who could benefit from nutritional intervention [36].

A pro-inflammatory status has been associated with several conditions that could affect the quality of life in the elderly, including depression and osteoporosis. In the next paragraphs, we focus on the role of DII in metabolic disorders, CVDs and cancer, as well as in healthy ageing and longevity.

5.3 DII and Metabolic Disorders

The metabolic syndrome is a complex disorder defined as a cluster of at least three risk factors among abdominal obesity, glucose intolerance, high blood pressure, high triglyceride levels and low high-density lipoprotein cholesterol levels [37].

Table 5.2 Relative risk (RR) and corresponding 95% confidence interval (CI) of metabolic syndrome for the highest versus the lowest category of DII score

Study	Study type	Country	Mean age	RR (95% CI)
NHANES [38]	Cohort	USA	46 years	1.23 (1.07–1.41)
SU.VI.MAX [39]	Cohort	France	49 years	1.39 (1.01–1.92)
SUN [40]	Cohort	Spain	35 years	0.86 (0.60–1.23)
KNHANES [28]	Case–control	Korea	41 years	1.67 (1.15–2.44)[a]
				1.40 (1.06–1.85)[b]
CUADHS [41]	Case–control	China	50 years	1.02 (0.75–1.40)
PONS [42]	Case–control	Poland	56 years	0.96 (0.77–1.19)
BCOPS [43]	Case–control	USA	42 years	0.87 (0.46–1.63)

[a]Postmenopausal women
[b]Men

Current evidence on the association between DII and metabolic syndrome is summarized in Table 5.2.

All studies but one were conducted in middle-aged individuals, reporting inconsistent results [44]. Two cohort studies reported a mild positive association, with an increased risk of metabolic syndrome in people with pro-inflammatory diets, whereas the SUN Project ("Seguimiento Universidad de Navarra" [University of Navarra Follow-up] Project is a dynamic prospective cohort study, conducted in Spain with university graduates since December 1999) found no association [44–46]. However, the younger age of patients in the SUN Project could have limited the capability to detect the effect of dietary indexes on the incidence of chronic diseases that are strongly related to age. Evidence from case–control studies generally indicates the lack of any association between DII and metabolic syndrome. Only one Korean case–control study found a significant association in both women (RR = 1.67; 95% CI: 1.15–2.44) and men (RR = 1.40; 95% CI: 1.06–1.85) [47]. Further, a cross-sectional study in overweight and sedentary individuals indicates a positive association between a metabolic syndrome score and pro-inflammatory diets [48].

Hyperglycaemia plays an important role in the definition of metabolic syndrome, and it has relevant implication on individual health. Generally, pro-inflammatory diets are associated with hyperglycaemia [45–47, 49, 50]. In particular, in the Mexico City Diabetes Mellitus Survey, a cross-sectional probabilistic population-based survey, individuals in the highest quintile of DII have a threefold higher risk of type 2 diabetes mellitus (95% CI: 1.39–6.58) than people in the lowest DII quintile [51]; this association was much stronger among people aged $\geq$55 years (OR = 9.8; 95% CI: 3.8–25.5) than among younger individuals. This result is in agreement with a previous case–control study in Iran, which reported an elevated risk of prediabetes (OR = 18.9; 95% CI: 7.0–50.8) in people taking pro-inflammatory diets [52]. In addition to the possible enhancement of insulin resistance in pro-inflammatory diets, elevated DII scores could identify diets rich in foods such as refined cereals, soft drinks, and red and processed meat associated with the onset of T2DM [51].

5.4 DII and Cardiovascular Diseases

The association between DII and CVDs has been investigated in several longitudinal studies on unselected populations (Table 5.3), generally reporting a moderate association. People taking pro-inflammatory diets reported an increased risk of CVD onset ranging from 3 to 103% compared with those taking anti-inflammatory diets. A similar risk was found for CVD mortality, with excess risk among people taking pro-inflammatory diets, ranging from 9 to 102%.

A few studies were conducted in the elderly. Only one study was specifically focused on women aged 70 years or older (mean age: 75 years) who were originally enrolled in the Calcium Intake Fracture Outcome Study [40]. Although this trial failed to demonstrate an effect of calcium supplementation on carotid atherosclerosis, it provided important information of dietary habits and CVD risk. In particular, pro-inflammatory diets were associated with increased mortality for atherosclerotic vascular disease (HR = 2.02; 95% CI: 1.30–3.13), ischaemic heart disease (HR = 2.51; 95% CI: 1.37–4.62) and possibly with cerebrovascular disease (HR = 1.76; 95% confidence interval 0.92–3.40). Three other studies enrolled participants with a mean age above 60 years. The PREDIMED Study, a Spanish long-term nutritional intervention study aimed at assessing the efficacy of the Mediterranean Diet in the primary prevention of cardiovascular disease, was a randomized trial enrolling men aged 55–80 years and women aged 60–80 years [39]. After a 5-year follow-up, a significant increase in CVD risk of 73% was observed in people following pro-inflammatory diets. This effect was consistent across dietary intervention arms. Two longitudinal studies investigated CVD mortality in women. In the

Table 5.3 Hazard ratio (HR) and corresponding 95% confidence interval (CI) of cardiovascular incidence and mortality for highest versus lowest category of DII score

Cohort	Country	Mean age	Mean follow-up	HR (95% CI)
Incidence				
MONICA/KORA [44]	Germany	55 years	23 years	1.53 (0.93–2.53)
ALSWH [45]	Australia	52 years	11 years	1.03 (0.76–1.42)
SU.VI.MAX [46]	France	49 years	11 years	1.15 (0.79–1.68)
GOS [47]	Australia	N.A.	5 years	2.00 (1.01–3.96)
SUN [48]	Spain	38 years	9 years	2.03 (1.06–3.88)
PREDIMED [35]	Spain	67 years	4 years	1.73 (1.15–2.60)
Mortality				
MCCS [16]	Australia	55 years	19 years	1.16 (1.01–1.33)
MEC [49]	US men	59 years	18 years	1.13 (1.03–1.23)
	US women			1.29 (1.17–1.42)
MONICA/KORA [44]	Germany	55 years	23 years	1.19 (0.76–1.86)
Whitehall II [12]	England	N.A.	22 years	1.46 (1.00–2.13)
CIFOS RCT [34]	Australia	75 years	13 years	2.02 (1.30–3.13)
NHANES III [50]	USA	47 years	14 years	1.46 (1.18–1.81)
IWHS [36]	USA	61 years	21 years	1.09 (1.01–1.18)
SMC [37]	Sweden	61 years	15 years	1.26 (0.93–1.70)

Iowa Women's Health Study designed to examine the effect of host, dietary and lifestyle factors on the incidence of cancer among postmenopausal women, a moderate association emerged between CVD mortality and DII score after adjustment for potential confounders [38]. Conversely, in the participants in the population-based Swedish Mammography Cohort, the association between DII and CVD mortality was no longer significant after adjustment for overweight/obesity, tobacco smoking and alcohol drinking [53].

Studies on different cardiovascular endpoints showed a consistent result. A recent meta-analysis of case–control and cohort studies revealed an increased risk of myocardial infarction in people with pro-inflammatory diets compared with those with anti-inflammatory diets (RR = 1.43; 95% CI: 1.09–1.89) [41]. Although not significant, the relative risks in people with elevated DII were still indicative of an excess risk for ischaemic/coronary heart disease (RR = 1.18; 95% CI: 0.89–1.58) and stroke (RR = 1.10; 95% CI: 0.60–2.00).

In summary, pro-inflammatory diets are associated with CVD incidence and mortality in both adults and the elderly. Therefore, the promotion of anti-inflammatory diets rich in vegetable, fish and cereals consumption may help to reduce CVD incidence, improving healthy ageing.

5.5   DII and Cancer

Several studies have shown that besides the well-known genetic and molecular alterations responsible for neoplastic transformation, also microenvironmental perturbations of the inflammatory status may lead to the accumulation of genetic damages responsible for malignant transformation of cells and to tumour progression [42, 43, 54]. A pro-inflammatory microenvironment, characterized by high concentrations of several pro-inflammatory cytokines and other mediators, is associated with an increased risk of tumour onset due to the inflammation-induced cellular stress and damages [55, 56].

However, alongside diet, the regulation of the cancer-related inflammatory status depends on other factors including ageing and the alteration of the organism redox state [18, 57]. Therefore, it is clear how all these factors may contribute to the formation of a more predisposing pro-inflammatory microenvironment and consequently to the development of cancer.

To date, there is a well-established axis among inflammation, diet, ageing and the increased risk of cancer, although the molecular basis of this multifactorial interaction has not yet been fully elucidated [58].

In the context of the diet-related pathogenesis of cancer, both quantitative and qualitative characteristics of nutrients play key roles in the development of a pro-inflammatory carcinogenic environment [59, 60]. Diet positively and negatively influences the incidence, natural progression and therapeutic response of several cancer types through the modulation of chronic inflammation [61, 62]. In particular, epidemiological studies demonstrate that certain types of cancers are more sensitive to disequilibria in food composition, especially when this imbalance characterizes obese people [63]. To date, among the most common malignancies associated with obesity or unhealthy dietary habits, more than 13 different types of cancers including cancer of the gallbladder, kidney cancer, liver cancer, breast cancer, ovarian cancer and thyroid cancer have been recognized [64].

In addition to overweight and obesity, specific foods and nutrients are now considered as probable or certain carcinogens for humans. The latest foods included in the International Agency for Research on Cancer list as probable and certain carcinogens for humans are the red and processed meats respectively [65]. In developed countries, the diet is generally rich in red meat, high-fat foods, refined grains and complex carbohydrates that promote a pro-inflammatory status [66].

Although in the Western Countries there is the highest percentage of overweight and obese people, in some areas, especially in those facing the Mediterranean Sea, healthy food styles are widespread, first of all being the Mediterranean diet, characterized by the consumption of high quantities of fruit, vegetables and olive oil [67]. The Mediterranean diet plays a protective role against various human diseases thanks to the various beneficial nutrients of which it is constituted [68]. One of the more important features of the Mediterranean diet is to have a low inflammatory index and antioxidant effects [33, 69]. Hence, the adoption of Mediterranean diet and healthy dietary habits may reduce the risk of cancer development and together the detrimental effects of ageing [70, 71].

Thanks to the use of the DII as a parameter for measuring the inflammatory potential of foods, several studies have been conducted to evaluate the role of pro-inflammatory diets on cancer risk. A recent meta-analysis summarized the current evidence based on case–control and cohort study [72]. Overall, a positive association between pro-inflammatory diets and overall cancer risk was found, with a RR = 1.58 (95% CI: 1.45–1.72) for the highest versus the lowest DII category. This overall risk estimate should be considered with caution, since it strongly depends on the cancer sites included in the original studies. Although a direct association emerged for all cancer sites, the risk magnitude was heterogeneous. Indeed, the pooled relative risks were 2.74 for oesophageal cancer, 1.57 for ovarian cancer, 1.45 for prostate cancer, 1.43 for colorectal cancer, 1.33 for breast cancer and 1.32 for renal cancer. The association was not significant for lung cancer (RR = 1.27; 95% CI: 0.93–1.72). Subgroup analysis by gender and ethnic group did not substantially modify study findings. A dose–response analysis generally showed a rising risk of cancer with increase DII score [72].

Considering that both DII score (Fig. 5.1) and overall cancer risk increase with age, it is compelling to understand whether their association changes according to age [73]. Using data from a series of case–control studies in Italy, Accardi et al. estimated the association between DII score and several cancer sites according to age groups [35]. The changes in DII scores according to age were similar in cancer cases and in cancer-free controls, so that the association between DII and relative risk for pro-inflammatory diets versus anti-inflammatory diets remained stable across age groups for all considered cancers. Indeed, the relative risk could be considered a magnifier of the baseline risk. Given the higher cancer incidence in the elderly, the impact of DII on absolute cancer risk was much higher among them than among middle-aged people [73]. Therefore, dietary intervention to improve healthy, anti-inflammatory diets in the elderly could help in reducing the cancer burden in this population.

Besides these epidemiological data, several animal models of cancer have also shown that the quantity of nutrients and the fasting and feeding cycles play an important role in cancer through their effects towards the immune system and the tumour microenvironment [60].

All these data suggest that dietary interventions may contribute to the regulation of both chronic low-grade metaflammation and inflamm-ageing. Therefore, diet is considered to play a key role not only in the pathogenesis and development of tumours but also as a possible therapeutic approach. This is based on the administration of specific foods or active natural compounds extracted from plants or microorganisms, i.e. nutraceuticals and probiotics, with proven antitumour activity or able to enhance the efficacy of the several pharmaceutical cancer treatments [74–77]. Finally, to date, diet is also considered a fundamental element in anticancer prevention strategies that promotes the adoption of healthy dietary lifestyles [68].

5.6   DII and Longevity

Several aspects may have an impact on healthy ageing, including good physical, cognitive and mental functioning. Therefore, conditions and lifestyle factors impacting on overall survival and on the incidence of chronic diseases and functional disabilities may be relevant.

Presently, very few studies have investigated the association between DII and healthy ageing using a multidimensional approach. Using data from the SU.VI. MAX, i.e. French Supplémentation en Vitamines et Minéraux Antioxydants study, the authors have prospectively evaluated the health status of almost 2800 participants (mean age at enrolment: 52 years) after a mean follow-up of 13.3 years [17]. Healthy ageing during follow-up was defined considering overall perceived health; physical and cognitive functioning; incidence of major chronic diseases; limitation in daily activities; depression and health-related limitations in social life. People having a pro-inflammatory diet at baseline had 15% lower probability to have healthy ageing than people with anti-inflammatory diets. Similarly, the Tsurugaya Project, a Japanese prospective study on community-dwelling older individuals aged 70 years or older (mean age: 75 years), evaluated the incidence of functional disabilities using a composite outcome. In this study, elevated DII was associated with a 25% increase in risk of developing disabilities during a maximum follow-up of 12 years [78].

Further, a Korean study on community-dwelling individuals aged 70–85 years investigated DII in relation to frailty [47]. A frailty index was calculated basing on five criteria: weight loss, exhaustion, low physical activity, low walking speed and low handgrip strength. A higher mean DII score was reported among frail individuals than among non-frail ones, with a 68% increase in the risk of frailty (95% CI: 1.25–2.17) among elderly with pro-inflammatory diets in comparison with anti-inflammatory diets. A similar association was found in a previous analysis of the Osteoarthritis Initiative [79]. In this study, frailty was defined according to the presence of at least two out of three of the following conditions: (1) weight loss ≥5%; (2) inability to carry out chair stand and (3) poor strength. During an 8-year follow-up, the incidence of frailty was higher in people taking pro-inflammatory diets than in those taking anti-inflammatory diets (HR = 1.37; 95% CI: 1.01–1.89).

Dietary habits may affect longevity through the modification of telomere length, which is considered a proxy of biological cell ageing [80] (Chap. 8). Chronic inflammation is a mechanism involved in telomere shortening; therefore, anti-inflammatory diets may contribute in slowing down biological ageing [81].

The association between leukocyte telomere length and inflammatory potential of diet was first investigated in the PREDIMED-NAVARRA, that is, in a subset of participants from the PREDIMED study recruited at NAVARRA centre [82]. The authors assessed the dietary inflammatory potential in 520 patients at high risk for CVD who underwent a nutritional intervention based on the Mediterranean diet. At study enrolment, the DII score was inversely associated with leukocyte telomere length at baseline, with an 80% higher risk of shorter telomere (95% CI: 1.03–3.17) in pro-inflammatory than in anti-inflammatory diets. More interestingly, pro-inflammatory diets were also inversely associated with telomere shortening rate after 5 years of follow-up: participants who had the greatest increase in DII had almost a twofold higher risk (OR = 1.94; 95% CI: 1.10–3.43) to have the highest telomere shortening rate.

These results should be considered with caution; although consistent with a beneficial effect of anti-inflammatory diets in slowing down telomere shortening, potential selection bias may limit their validity. Indeed, the PREDIMED-NAVARRA study was based on a population with several comorbidities (i.e., diabetes, overweight/obesity, hypertension and current smoking), which may confound the reported association between DII and telomere length. Although no interaction was found, there was a tendency for individuals with overweight/obesity or hypertension to have shorter telomeres [82].

A subsequent analysis of data from the NHANES study, a programme designed to assess the health and nutritional status of children and adults in the United States, confirmed an inverse association between DII and telomere length, with shorter leukocyte telomeres in people with the most pro-inflammatory DII scores [83]. This analysis was conducted on a probabilistic sample of US general population of over 7000 individuals aged 18 years or older. Conversely, no such association was found in middle-aged Asklepios population, i.e. a community situated on both sides of the busy motorway from Brussels to Ghent, with divergent effects in men and women [84].

People adherent to Mediterranean diet reported anti-inflammatory scores [34, 35]; therefore, an indirect support to the association between pro-inflammatory diets and telomere shortening may derive from study on Mediterranean diet. Although presently scarce, the current evidence confirms an inverse association between telomere length and adherence to the Mediterranean diet [85]. This association was not evident for any single Mediterranean diet food, suggesting that the dietary Mediterranean pattern, rather than single foods or nutrients, may be responsible for the beneficial effect on telomere shortening. An Italian study on Caucasian elderly reported a similar association between adherence to Mediterranean diet and telomere length in people over 70 years of age [86].

Finally, in two large prospective Spanish cohorts, the SUN [87] and PREDIMED studies [39], a high inflammatory diet, as measured by the DII, was associated with higher all-cause mortality. A diet rich in vegetables, fruits, fish, nuts and legumes and low in meats, dairy and baked goods, that is, a higher anti-inflammatory DII, is therefore likely to reduce many potential causes of premature death, potentially favouring the attainment of longevity [88].

5.7   Conclusion

The complexity of the ageing and longevity phenomena and their regulatory elements is progressively being revealed thanks to the multidisciplinary studies carried out in recent years in this field. The predominant role of inflammation in these physiological processes has made possible to understand how the accurate regulation of the inflammatory status is responsible for the development of different pathological processes, shifting the interest of researchers on the factors able to change inflammatory status. As described in this chapter, several studies have shown that diet is one of the most effective regulators of inflammatory processes by identifying specific pro-inflammatory and anti-inflammatory dietary habits associated with different pathologies or their treatment. An important turning point in the study of diet-associated diseases was the definition of the DII, which allows to quantify the pro- or anti-inflammatory power of a specific food. After the introduction of the DII, it is possible to identify dietary regimens associated with an increased risk of metabolic disorders, CVDs and cancer and reciprocally healthy ageing and longevity. It is becoming clear that the comprehensive study of inflammation, diet, ageing and longevity and the understanding of the relationships between these four elements can provide useful information to predict the risk of occurrence of specific diseases. This will be helpful to develop a new preventive and therapeutic approach based on the modulation of diet, thereby enhancing the possibility to achieve healthy ageing and longevity.

References

1. Leon BM, Maddox TM. Diabetes and cardiovascular disease: epidemiology, biological mechanisms, treatment recommendations and future research. World J Diabetes. 2015;6(13):1246–58.
2. Koene RJ, Prizment AE, Blaes A, Konety SH. Shared risk factors in cardiovascular disease and cancer. Circulation. 2016;133(11):1104–14.
3. Rose MR, Flatt T, Graves JL, Greer LF, Martinez DE, Matos M, et al. What is aging? Front Genet. 2012;3:134.
4. Ahmed AS, Sheng MH, Wasnik S, Baylink DJ, Lau KW. Effect of aging on stem cells. World J Exp Med. 2017;7(1):1–10.

5. López-Otín C, Blasco MA, Partridge L, Serrano M, Kroemer G. The hallmarks of aging. Cell. 2013;153(6):1194–217.
6. Leonardi GC, Accardi G, Monastero R, Nicoletti F, Libra M. Ageing: from inflammation to cancer. Immun Ageing. 2018;15:1.
7. Accardi G, Caruso C. Immune-inflammatory responses in the elderly: an update. Immun Ageing. 2018;15:11.
8. Hunter GR, Gower BA, Kane BL. Age related shift in visceral fat. Int J Body Compos Res. 2010;8(3):103–8.
9. Deng T, Lyon CJ, Bergin S, Caligiuri MA, Hsueh WA. Obesity, inflammation, and cancer. Annu Rev Pathol. 2016;11:421–49.
10. Suiter C, Singha SK, Khalili R, Shariat-Madar Z. Free fatty acids: circulating contributors of metabolic syndrome. Cardiovasc Hematol Agents Med Chem. 2018;16(1):20–34.
11. Ellulu MS, Patimah I, Khaza'ai H, Rahmat A, Abed Y. Obesity and inflammation: the linking mechanism and the complications. Arch Med Sci. 2017;13(4):851–63.
12. Forsythe LK, Wallace JM, Livingstone MB. Obesity and inflammation: the effects of weight loss. Nutr Res Rev. 2008;21(2):117–33.
13. Gregor MF, Hotamisligil GS. Inflammatory mechanisms in obesity. Annu Rev Immunol. 2011;29:415–45.
14. Xia S, Zhang X, Zheng S, Khanabdali R, Kalionis B, Wu J, et al. An update on inflamm-aging: mechanisms, prevention, and treatment. J Immunol Res. 2016;2016:8426874.
15. Frasca D, Blomberg BB, Paganelli R. Aging, obesity, and inflammatory age-related diseases. Front Immunol. 2017;8:1745.
16. Bottazzi B, Riboli E, Mantovani A. Aging, inflammation and cancer. Semin Immunol. 2018;40:74–82.
17. Assmann KE, Adjibade M, Shivappa N, Hébert JR, Wirth MD, Touvier M, et al. The inflammatory potential of diet at midlife is associated with later healthy aging in French adults. J Nutr. 2018;148(3):437–44.
18. Minihane AM, Vinoy S, Russell WR, Baka A, Roche HM, Tuohy KM, et al. Low-grade inflammation, diet composition and health: current research evidence and its translation. Br J Nutr. 2015;114(7):999–1012.
19. Shivappa N, Steck SE, Hurley TG, Hussey JR, Hebert JR. Designing and developing a literature-derived, population-based dietary inflammatory index. Public Health Nutr. 2014;17(8):1689–96.
20. Cătană CS, Atanasov AG, Berindan-Neagoe I. Natural products with anti-aging potential: affected targets and molecular mechanisms. Biotechnol Adv. 2018;36(6):1649–56.
21. Muñoz A, Costa M. Nutritionally mediated oxidative stress and inflammation. Oxid Med Cell Longev. 2013;2013:610950.
22. Tan BL, Norhaizan ME, Liew WP. Nutrients and oxidative stress: friend or foe? Oxid Med Cell Longev. 2018;2018:9719584.
23. Zhang X, Li J, Sejas DP, Pang Q. The ATM/p53/p21 pathway influences cell fate decision between apoptosis and senescence in reoxygenated hematopoietic progenitor cells. J Biol Chem. 2005;280(20):19635–40.
24. Sasaki M, Ikeda H, Sato Y, Nakanuma Y. Proinflammatory cytokine-induced cellular senescence of biliary epithelial cells is mediated via oxidative stress and activation of ATM pathway: a culture study. Free Radic Res. 2008;42(7):625–32.
25. Bartek J, Hodny Z, Lukas J. Cytokine loops driving senescence. Nat Cell Biol. 2008;10(8):887–9.
26. Soria-Valles C, López-Soto A, Osorio FG, López-Otín C. Immune and inflammatory responses to DNA damage in cancer and aging. Mech Ageing Dev. 2017;165(Pt A):10–6.
27. Shivappa N, Hebert JR, Rietzschel ER, De Buyzere ML, Langlois M, Debruyne E, et al. Associations between dietary inflammatory index and inflammatory markers in the Asklepios study. Br J Nutr. 2015;113(4):665–71.
28. Tabung FK, Steck SE, Zhang J, Ma Y, Liese AD, Agalliu I, et al. Construct validation of the dietary inflammatory index among postmenopausal women. Ann Epidemiol. 2015;25(6):398–405.

29. Shivappa N, Hebert JR, Marcos A, Diaz LE, Gomez S, Nova E, et al. Association between dietary inflammatory index and inflammatory markers in the HELENA study. Mol Nutr Food Res. 2017;61(6):1600707.
30. Shivappa N, Hebert JR, Kivimaki M, Akbaraly T. Alternate healthy eating index 2010, dietary inflammatory index and risk of mortality: results from the Whitehall II cohort study and meta-analysis of previous dietary inflammatory index and mortality studies. Br J Nutr. 2017;118(3):210–21.
31. Alkerwi A, Vernier C, Crichton GE, Sauvageot N, Shivappa N, Hébert JR. Cross-comparison of diet quality indices for predicting chronic disease risk: findings from the observation of cardio-vascular risk factors in Luxembourg (ORISCAV-LUX) study. Br J Nutr. 2015;113(2):259–69.
32. Wirth MD, Hébert JR, Shivappa N, Hand GA, Hurley TG, Drenowatz C, et al. Anti-inflammatory dietary inflammatory index scores are associated with healthier scores on other dietary indices. Nutr Res. 2016;36(3):214–9.
33. Billingsley HE, Carbone S. The antioxidant potential of the Mediterranean diet in patients at high cardiovascular risk: an in-depth review of the PREDIMED. Nutr Diabetes. 2018;8(1):13.
34. Hodge AM, Bassett JK, Dugué PA, Shivappa N, Hébert JR, Milne RL, et al. Dietary inflam-matory index or Mediterranean diet score as risk factors for total and cardiovascular mortality. Nutr Metab Cardiovasc Dis. 2018;28(5):461–9.
35. Accardi G, Shivappa N, Di Maso M, Hébert JR, Fratino L, Montella M, et al. Dietary inflam-matory index and cancer risk in elderly: a pooled-analysis of Italian case-control studies. Nutrition. 2019;63–64:205–10.
36. Hernández Morante JJ, Martínez CG, Morillas-Ruiz JM. Dietary factors associated with frailty in old adults: a review of nutritional interventions to prevent frailty development. Nutrients. 2019;11(1):E102.
37. Kassi E, Pervanidou P, Kaltsas G, Chrousos G. Metabolic syndrome: definition and controver-sies. BMC Med. 2011;9:48.
38. Shivappa N, Blair CK, Prizment AE, Jacobs DR Jr, Steck SE, Hébert JR. Association between inflammatory potential of diet and mortality in the Iowa Women's health study. Eur J Nutr. 2016;55(4):1491–502.
39. Garcia-Arellano A, Ramallal R, Ruiz-Canela M, Salas-Salvadó J, Corella D, Shivappa N, et al. Dietary inflammatory index and incidence of cardiovascular disease in the PREDIMED study. Nutrients. 2015;7(6):4124–38.
40. Bondonno NP, Lewis JR, Blekkenhorst LC, Shivappa N, Woodman RJ, Bondonno CP, et al. Dietary inflammatory index in relation to sub-clinical atherosclerosis and atherosclerotic vas-cular disease mortality in older women. Br J Nutr. 2017;117(11):1577–86.
41. Shivappa N, Godos J, Hébert JR, Wirth MD, Piuri G, Speciani AF, et al. Dietary inflammatory index and cardiovascular risk and mortality—a meta-analysis. Nutrients. 2018;10(2):E200.
42. Kawanishi S, Ohnishi S, Ma N, Hiraku Y, Murata M. Crosstalk between DNA damage and inflammation in the multiple steps of carcinogenesis. Int J Mol Sci. 2017;18(8):E1808.
43. Sui X, Lei L, Chen L, Xie T, Li X. Inflammatory microenvironment in the initiation and pro-gression of bladder cancer. Oncotarget. 2017;8(54):93279–94.
44. Pimenta AM, Toledo E, Rodriguez-Diez MC, Gea A, Lopez-Iracheta R, Shivappa N, et al. Dietary indexes, food patterns and incidence of metabolic syndrome in a Mediterranean cohort: the SUN project. Clin Nutr. 2015;34(3):508–14.
45. Neufcourt L, Assmann KE, Fezeu LK, Touvier M, Graffouillère L, Shivappa N, et al. Prospective association between the dietary inflammatory index and metabolic syndrome: findings from the SU.VI.MAX study. Nutr Metab Cardiovasc Dis. 2015;25(11):988–96.
46. Mazidi M, Shivappa N, Wirth MD, Hebert JR, Mikhailidis DP, Kengne AP, et al. Dietary inflammatory index and cardiometabolic risk in US adults. Atherosclerosis. 2018;276:23–7.
47. Kim HY, Lee J, Kim J. Association between dietary inflammatory index and metabolic syn-drome in the general Korean population. Nutrients. 2018;10(5):E648.
48. Camargo-Ramos CM, Correa-Bautista JE, Correa-Rodríguez M, Ramírez-Vélez R. Dietary inflammatory index and cardiometabolic risk parameters in overweight and sedentary subjects. Int J Environ Res Public Health. 2017;14(10):E1104.

49. Ren Z, Zhao A, Wang Y, Meng L, Szeto IMY, Li T, et al. Association between dietary inflammatory index, C-reactive protein and metabolic syndrome: a cross-sectional study. Nutrients. 2018;10(7):E831.
50. Sokol A, Wirth MD, Manczuk M, Shivappa N, Zatonska K, Hurley TG, et al. Association between the dietary inflammatory index, waist-to-hip ratio and metabolic syndrome. Nutr Res. 2016;36(11):1298–303.
51. Denova-Gutiérrez E, Muñoz-Aguirre P, Shivappa N, Hébert JR, Tolentino-Mayo L, Batis C, et al. Dietary inflammatory index and type 2 diabetes mellitus in adults: the diabetes mellitus survey of Mexico City. Nutrients. 2018;10(4):382.
52. Vahid F, Shivappa N, Karamati M, Naeini AJ, Hébert JR, Davoodi SH. Association between dietary inflammatory index (DII) and risk of prediabetes: a case-control study. Appl Physiol Nutr Metab. 2017;42(4):399–402.
53. Shivappa N, Harris H, Wolk A, Héberty JR. Association between inflammatory potential of diet and mortality among women in the Swedish mammography cohort. Eur J Nutr. 2016;55(5):1891–900.
54. Mantovani A, Allavena P, Sica A, Balkwill F. Cancer-related inflammation. Nature. 2008;454(7203):436–44. https://doi.org/10.1038/nature07205.
55. Multhoff G, Molls M, Radons J. Chronic inflammation in cancer development. Front Immunol. 2011;2:98.
56. Landskron G, De la Fuente M, Thuwajit P, Thuwajit C, Hermoso MA. Chronic inflammation and cytokines in the tumor microenvironment. J Immunol Res. 2014;2014:149185.
57. Zinger A, Cho WC, Ben-Yehuda A. Cancer and aging—the inflammatory connection. Aging Dis. 2017;8(5):611–27.
58. Sapienza C, Issa JP. Diet, nutrition, and cancer epigenetics. Annu Rev Nutr. 2016;36:665–81.
59. Mierke CT. The fundamental role of mechanical properties in the progression of cancer disease and inflammation. Rep Prog Phys. 2014;77(7):076602.
60. Zitvogel L, Pietrocola F, Kroemer G. Nutrition, inflammation and cancer. Nat Immunol. 2017;18(8):843–50.
61. Thomas F, Rome S, Mery F, Dawson E, Montagne J, Biro PA, et al. Changes in diet associated with cancer: an evolutionary perspective. Evol Appl. 2017;10(7):651–7.
62. Soldati L, Di Renzo L, Jirillo E, Ascierto PA, Marincola FM, De Lorenzo A. The influence of diet on anti-cancer immune responsiveness. J Transl Med. 2018;16(1):75.
63. Mazzarella L. Why does obesity promote cancer? Epidemiology, biology, and open questions. Ecancermedicalscience. 2015;9:554.
64. Hooper L, Anderson AS, Birch J, Forster AS, Rosenberg G, Bauld L, et al. Public awareness and healthcare professional advice for obesity as a risk factor for cancer in the UK: a cross-sectional survey. J Public Health. 2018;40(4):797–805.
65. IARC Working Group on the Evaluation of Carcinogenic Risk to Humans. Red meat and processed meat. Lyon: International Agency for Research on Cancer; 2018.
66. Smidowicz A, Regula J. Effect of nutritional status and dietary patterns on human serum C-reactive protein and interleukin-6 concentrations. Adv Nutr. 2015;6(6):738–47.
67. Vasto S, Buscemi S, Barera A, Di Carlo M, Accardi G, Caruso C. Mediterranean diet and healthy ageing: a Sicilian perspective. Gerontology. 2014;60(6):508–18.
68. Romagnolo DF, Selmin OI. Mediterranean diet and prevention of chronic diseases. Nutr Today. 2017;52(5):208–22.
69. Casas R, Sacanella E, Estruch R. The immune protective effect of the Mediterranean diet against chronic low-grade inflammatory diseases. Endocr Metab Immune Disord Drug Targets. 2014;14(4):245–54.
70. Martucci M, Ostan R, Biondi F, Bellavista E, Fabbri C, Berżarelli C, et al. Mediterranean diet and inflammaging within the hormesis paradigm. Nutr Rev. 2017;75(6):442–55.
71. Schwingshackl L, Schwedhelm C, Galbete C, Hoffmann G. Adherence to Mediterranean diet and risk of cancer: an updated systematic review and meta-analysis. Nutrients. 2017;9(10):E1063.

72. Li D, Hao X, Li J, Wu Z, Chen S, Lin J, et al. Dose-response relation between inflammatory index and human cancer risk: evidence from 44 epidemiological studies involving 1,082,092 participants. Am J Clin Nutr. 2018;107(3):371–88.
73. Bray F, Colombet M, Mery L, Piñeros M, Znaor A, Zanetti R and Ferlay J. editors. Cancer incidence in five continents, Vol. XI (electronic version). Lyon: International Agency for Research on Cancer; 2017. http://ci5.iarc.fr. Accessed 14 Jan 2019.
74. Banna GL, Torino F, Marletta F, Santagati M, Salemi R, Cannarozzo E, et al. *Lactobacillus rhamnosus* GG: an overview to explore the rationale of its use in cancer. Front Pharmacol. 2017;8:603.
75. Hamzehzadeh L, Atkin SL, Majeed M, Butler AE, Sahebkar A. The versatile role of curcumin in cancer prevention and treatment: a focus on PI3K/AKT pathway. J Cell Physiol. 2018;233(10):6530–7.
76. Vivarelli S, Salemi R, Candido S, Falzone L, Santagati M, Stefani S, et al. Gut microbiota and cancer: from pathogenesis to therapy. Cancers (Basel). 2019;11(1). pii: E38.
77. Falzone L, Salomone S, Libra M. Evolution of cancer pharmacological treatments at the turn of the third millennium. Front Pharmacol. 2018;9:1300.
78. Tomata Y, Shivappa N, Zhang S, Nurrika D, Tanji F, Sugawara Y, et al. Dietary inflammatory index and disability-free survival in community-dwelling older adults. Nutrients. 2018;10(12):E1896.
79. Shivappa N, Stubbs B, Hébert JR, Cesari M, Schofield P, Soysal P, et al. The relationship between the dietary inflammatory index and incident frailty: a longitudinal cohort study. J Am Med Dir Assoc. 2018;19(1):77–82.
80. Müezzinler A, Zaineddin AK, Brenner H. A systematic review of leukocyte telomere length and age in adults. Ageing Res Rev. 2013;12(2):509–19.
81. Wong JY, De Vivo I, Lin X, Fang SC, Christiani DC. The relationship between inflammatory biomarkers and telomere length in an occupational prospective cohort study. PLoS One. 2014;9(1):e87348.
82. García-Calzón S, Zalba G, Ruiz-Canela M, Shivappa N, Hébert JR, Martínez JA, et al. Dietary inflammatory index and telomere length in subjects with a high cardiovascular disease risk from the PREDIMED-NAVARRA study: cross-sectional and longitudinal analyses over 5 y. Am J Clin Nutr. 2015;102(4):897–904.
83. Shivappa N, Wirth MD, Hurley TG, Hébert JR. Association between the dietary inflammatory index (DII) and telomere length and C-reactive protein from the National Health and Nutrition Examination Survey-1999-2002. Mol Nutr Food Res. 2017;61(4):1600630.
84. De Meyer T, Bekaert S, De Buyzere ML, De Bacquer DD, Langlois MR, Shivappa N, et al. Leukocyte telomere length and diet in the apparently healthy, middle-aged Asklepios population. Sci Rep. 2018;8(1):6540.
85. Freitas-Simoes TM, Ros E, Sala-Vila A. Nutrients, foods, dietary patterns and telomere length: update of epidemiological studies and randomized trials. Metabolism. 2016;65(4):406–15.
86. Boccardi V, Esposito A, Rizzo MR, Marfella R, Barbieri M, Paolisso G. Mediterranean diet, telomere maintenance and health status among elderly. PLoS One. 2013;8(4):e62781.
87. Ramallal R, Toledo E, Martínez-González MA, Hernández-Hernández A, García-Arellano A, et al. Dietary inflammatory index and incidence of cardiovascular disease in the SUN cohort. PLoS One. 2015;10(9):e0135221.
88. Garcia-Arellano A, Martínez-González MA, Ramallal R, Salas-Salvadó J, Hébert JR, Corella D, eSUN and PREDIMED Study Investigators. Dietary inflammatory index and all-causemortality in large cohorts: the SUN and PREDIMED studies. Clin Nutr. 2019;38(3):1221–31.

Genetic Signatures of Centenarians

6

Francesco Villa, Anna Ferrario,
and Annibale Alessandro Puca

6.1   Exceptional Longevity Is an Inherited Phenotype

Centenarians are individuals who live almost 50 years more than their cohort; if we consider that at the beginning of the twentieth century, life expectancy was around 50 years [1]. They delay or escape from the diseases of ageing, which are mostly cardiovascular. Exceptional longevity is a complex phenotype conditioned by many factors that, in some way, perturb organism during the whole life. Physical activity, pollution, diet, chronic stress, psychological disorders and smoking are the major external factors that influence the ageing process. However, all these players act with different impact on different individuals because of the unique genetic background that strongly characterizes people's attitude to live longer and healthier. Many studies, indeed, showed that human genome hides the secrets to face successfully the age-related diseases or to escape them, living the old age in healthy condition. Indeed, one study showed that exceptional longevity phenotype runs in family, calculating that the probability to find such clusters by chance is less than 1 in a billion [2, 3]. Then, the relative risk of becoming a centenarian for a centenarian sibling was several folds higher than that in the general population, and the mortality across the different ages was one half. Furthermore, offspring of centenarians have reduced mortality and are less exposed to cardiovascular diseases [1, 3].

F. Villa · A. Ferrario
Cardiovascular Research Unit, IRCCS MultiMedica, Milan, Italy
e-mail: francesco.villa@multimedica.it; anna.ferrario@multimedica.it

A. A. Puca (✉)
Cardiovascular Research Unit, IRCCS MultiMedica, Milan, Italy

Department of Medicine, Surgery and Dentistry "Scuola Medica Salernitana", University of Salerno, Baronissi, SA, Italy
e-mail: apuca@unisa.it

6.2   Case–Control Candidate Gene Approach

The case–control study is a retrospective analysis that starts from two different groups: one with a disease/outcome and one without it. The aim of the study is the assessment of the presence of significant difference in the rate of exposure to a given risk factor between the two groups.

It was initially adopted to investigate the hypothesis of a possible involvement of genes based on the a priori knowledge of their physiological, biological or biochemical function (candidate gene approach).

The complexity of the ageing process is determined by the participation of several factors, either external intervention, such as diet and lifestyle that interfere with epigenetic mechanisms, [1] or genetic variations (single nucleotide polymorphisms or SNPs, deletions, duplications, etc.) that affect players involved in important biological pathways. The candidate gene approach, speculating on the involvement of specific genes, aims to identify the possible causative variants that impact their function. Examples of this approach is the test of genetic variants in genes involved in lipid metabolism (apolipoprotein E or APOE, apolipoprotein B or APOB), insulin signalling (Forkhead box O3A or FOXO3A), immunology (Toll-like receptor 4 or TLR4), radical stress buffering (Paraoxonase 1 or PON1), ion trafficking [4–7]. After many attempts and associations published on the many candidate genes tested, only few variants have been consistently associated with exceptional longevity due to inconsistent replication observed in independent cohorts. There are a lot of reasons for replication failure including lack of statistical power, the generation of false positives due to population stratification, differences in adopted criteria for inclusion among studies and different genetic background and environmental factors among analysed populations [2]. In detail, a result is reliable if supported by an adequate number of study participants that is calculated based on the number of hypothesis tested and on the frequency of the genetic variant analysed. Indeed, more hypothesis and rare variants require larger populations then one hypothesis and a common variant [2]. Furthermore, if we compare affected individuals (case) with unaffected individuals (controls) who are not balanced in terms of genetic background, the observed skewing is probably due to the differences in frequency unrelated to the phenotype but depending on the genetic background, the so-called genetic admixture. This is now clear, thanks to the massive genotyping (SNP array and whole genome sequencing) that allows to reduce the genetic admixture through a genetic component analysis aimed to exclude possible genetic outliers [8]. Another important aspect is the age and gender of the populations recruited for the study. Some studies adopted very relaxed threshold for inclusion criteria, such as age. A 95-year-old male is not equivalent to a 95-year-old female, being that males live 5 years less than females on average, so age inclusion criteria must differ between males and females, especially if relaxed threshold is used. The lack of such differentiation has generated different demographic pressures, i.e., the difficulties encountered during a life that allow some to survive if carrier of a protective genetic background, with males more selected by a more severe age cut-off than females if the same age is adopted. Indeed, there are gender-specific associations in male, as for FOXO3A in the Southern Italian Centenarian Study (SICS) [9]. Finally, the demographic pressure can change based on different environments. Let us think about altitude, for example. People exposed to altitudes will survive better if carriers of polymorphisms, affecting the ability to adapt to low oxygen levels, and this could not be true for people who live by the sea. Notably, most of the long-living individuals of the SICS, enrolled in the Italian region of Cilento, live on the mountains (personal communication by Dr. Giuseppina Arcaro). Another example of the environment's importance could be the presence of epidemic infectious diseases, making the efficiency of the immunological system crucial for survival. So, comparison of different studies is also very challenging, as shown by the different meta-analyses performed (see below).

Thus, of the many genes tested, only ApoE and FOXO3A survived to association in independent populations. ApoE is the principal cholesterol carrier that drives lipid transport and injury repair in the brain. The ApoE gene is polymorphic, and the study of Garatachea et al. [10], focused on the three most common alleles ε2, ε3, and ε4, showed that extreme longevity was negatively associated with ε4 allele carriage, while the presence of the ε2/ε3 genotype was positively associated with it. ApoE ε4 is correlated with increased risk of developing age-related diseases such as cardiovascular disease and Alzheimer's disease. Indeed, ApoE knockout mice develops atherosclerosis if fed with a high-fat diet, further strengthening that genes involved in the homeostasis of vascular system could be a potential candidate for genetic association studies in exceptional longevity.

The FOXO3A rs2802292 is another variant found associated with exceptional longevity across many populations (but not all) though the polymorphism has not a clear function. Nevertheless, FOXO3A is part of the longevity pathway IGF1/PI3K/PDK1/AKT/FOXO and has an important role as 'gatekeeper' by balancing the cell response to oxidative stress and nutrient availability. The polymorphism may improve the ability of FOXO3A in fighting oxidative stress by enhancing its interconnections with up- and downstream molecular partners. This axis is conserved and associated with longevity in different species from worms to humans [11]. Many aspects of cellular homeostasis, survival, proliferation and oxidative stress response are regulated by this pathway [12].

Recently, a study conducted in collaboration with University of Palermo and Medical University of South Carolina [13] showed that GM3 allotype, that is one of the hereditary antigenic determinants expressed on immunoglobulin polypeptide heavy chains of IgG1, is overrepresented in a subset of long-living individuals (LLIs) enrolled in the SICS, assessing the role of the genetic component of immune response players in longevity. The GM3 allotype replaces the lysine of the GM17 with an arginine in the Constant Heavy 1 (CH1) region. The study was conducted using sequencing technology (see below) on a relatively small number of samples, and it needs to be replicated, but it is the first report indicating GM allotypes among genetic signatures of the centenarian trait. Notably, this locus is not covered by commercial SNP arrays (see below).

6.3   Case–Control Studies Without A Priori Hypothesis (Non-hypothesis Driven)

An initial use of the case–control study design to interrogate genes without a priori hypothesis based on the gene function was run by Geesaman et al. [14] who reported a discovery of SNPs in the 4q25 chromosome locus interrogated under a case–control study design that was a follow-up of the genome-wide linkage scan on more than 300 familial extremely long-lived [15]. A haplotype block map (see below) was built on the entire region based on the polymorphisms that were found and the association study performed on these haplotypes revealed a polymorphism in the promoter of microsomal transfer protein (MTP) gene, already described as a gene expression modulator, as a modifier of human lifespan. MTP is a lipid transfer protein required for the assembly and secretion of lipoprotein and is directly involved in the packaging of chylomicrons composed of ApoB and triglycerides. MTP is involved in a rate-limiting step in lipid metabolism and thus correlated with human longevity: coronary artery disease and other vasculopathies are attributed to adverse lipid profile and a variant that impacts the function of lipid metabolism could affect human lifespan. MTP was not replicated in other studies, but a possible explanation was generated by the group of Nir Barzilai, which showed a buffering effect by other variants, that is, other polymorphisms could influence the protective effects of an allele [16].

Thanks to the availability of platforms able to interrogate hundreds of thousands of SNPs by hybridizing genomic DNA on a chip array, association studies could be done at the genomic level without peaking a candidate gene. By doing so, case–control studies at the genome level (genome-wide association studies or GWAS) generated an enormous amount of data on almost all the variants that characterize almost all the human genes. Indeed, it was clear that by analysing a smaller number of SNPs, a much broader number of variants are interrogated thanks to the linkage disequilibrium, that is, most of the SNPs were inherited together with a small level of independence, the haplotype blocks [17, 18]. Thus, millions of SNPs could be analysed by interrogating few hundreds of thousands of them across the genome and then associated with the phenotype of interest.

While GWAS are hypothesis-free attempts, they generate an enormous number of tests, reducing the power of the study and forcing to adopt very low levels of threshold of significance ($p < 10^{-8}$).

GWAS on exceptional longevity produced important discoveries on the role of genetic risk factors for diseases of ageing. Indeed, many variations have been identified, but only few are involved in age-related diseases. The association of APOE that emerged from the vast majority of studies is a remarkable example. The analysed cohort in replication studies, which are mandatory to confirm or discard a genetic association, is not homogeneous, differing in terms of gender and age of the individuals.

Among the recent studies, Sebastiani et al. in 2012 published a study run on a discovery cohort of more than 800 centenarians and 900 genetically matched controls and two replication cohorts for an additional 300 centenarians and 3200 controls. Using a 370k chip of Illumina, Sebastiani and colleagues generated a list of SNPs potentially associated with exceptional longevity and used the most informative SNPs to build signatures that were able to predict the survival and mortality of the carriers [19]. In particular, the study revealed the presence of different clusters in populations with similar genetic characteristics. Using a Bayesian classification model, that is, a statistical classification based on the Bayes' theorem that can predict the probability that an object belongs to a particular subset, 281 SNPs, of which 131 sited in 130 genes, have been identified as potentially associated with extreme longevity trait. Only few of them are in genes that have been previously associated with longevity in other studies. One of them is Lamin A/C gene (LMNA) that is known for its role in Hutchinson-Gilford progeria syndrome (HGPS), which causes a severe form of premature ageing due to the intracellular accumulation of mutated protein that, in turn, causes DNA damages. In an extensive meta-analysis, two SNPs, belonging to a conserved four-SNP haplotype, in LMNA have been associated with longevity [20]. The Werner Syndrome RecQ Like Helicase gene (WRN), responsible for the atherosclerotic progeroid syndrome with the same name, has also been associated with longevity through two common variants, which are able to lower the degree of coronary artery occlusion [21] Moreover, superoxide dismutase 2 (SOD2) has been extensively analysed for its role in oxidative stress regulation and has been associated with longevity in a Danish centenarian cohort [22]. The SNP with the highest association p-value was found to be rs2075650 in APOE. The 130 genes have been computationally annotated, and the analysis showed that 30 of them are associated with Alzheimer disease (AD), 24 with coronary artery disease (CAD), 42 with dementia and 38 with tauopathies in literature [19]. Epidemiologically, these results reflect the fact that in LLIs, these pathologies are almost absent and the age of onset is extremely delayed [23].

Villa et al., in 2015, published a concordant association in BPIFB4 (rs2070325) in three independent populations of long-living individuals. This SNP was part of a 4-SNP haplotype that codifies for a Longevity Associated Variant of BPIFB4 (LAV-BPIFB4), a protein that is able to activate endothelial nitric oxide synthase (eNOS) through calcium mobilization, protein kinase C alpha (PKCα) activation and the recruitment of heat shock protein (HSP) 90 and 14-3-3 scaffold protein [24–26]. It is well known that nitric oxide (NO) plays a fundamental role in healthy ageing of endothelium, and its impairment is associated with vessel frailty and ageing [27]. Therapy with LAV-BPIFB4 induced a reduction in blood pressure in hypertensive rats, a rescue of endothelial dysfunction in old mice and revascularization in a murine model of limb ischemia [24]. Furthermore, BPIFB4 is more abundant in sera of LLIs as compared to young controls and in healthy LLIs as compared to frail ones [28]. This was also true at the mRNA level in mononuclear cells (MNCs) of LLIs versus young controls, and the up-regulation was concomitant with CXCR4 mRNA down-regulation and its activation by membrane recruitment [29]. A rare variant (RV) of BPIFB4, on the other hand, induced high blood pressure and endothelial dysfunction in mice, which were associated with high blood pressure in patients under therapy. It is plausible that the LAV haplotype, which has been selected by evolutionary forces, has been the object of a recombination event generating a chimera with only SNPs in positions 3 and 4 of the haplotype mutated generating a disadvantageous haplotype [30]. These results underscore the importance of functional studies to validate genetic association studies.

6.4   Meta-Analysis of GWAS

The need for the adequate statistical power in GWAS brought researchers to create models of statistical analyses for combining different studies with the aim to enlarge analysed populations. A meta-analysis is not just a sum of samples. Each population is characterized by many factors that can include important biases for the final results. First, the age of the population: studies include groups of different mean ages because of either the characteristic of the population or the selected age cut-off. This cut-off is not universally set up because of the complexity of the trait that suffers the influx of many stimuli and can be very different between populations. Some studies adopted very relaxed age threshold (>90 years), while others were more stringent (>100 years). However, in the last 10 years, the majority of the studies used cut-off growing from 90 towards 100 years of age.

Another important issue is the genetic admixture of the populations (see above). To evaluate the ethnic homogeneity of the sample populations, in addition to the genetic component (GC) analysis that allows excluding the outliers from the study, allele and genotype frequencies in the absence of other evolutionary influences must be in equilibrium. This equilibrium, called Hardy-Weinberg Equilibrium (HWE), must be maintained by populations involved in the meta-analysis.

Starting from the above observations, Sebastiani et al. adopted a very stringent threshold for age cut-off criteria of inclusion (<1 percentile of their cohort) for a meta-analysis of four GWAS [31]. The results, based on a smaller but older set as compared to study cohorts adopted, generated new findings including a rare variant in chromosome 4q25 on Elongation Of Long Chain Fatty Acids 6 gene (ELOVL6). This is of particular interest due to the linkage results already described and the ELOVL6 function in modulating palmitoleic acid levels, which have been found to be correlated with longevity from worms to humans [32]. Palmitoleic acid is a monounsaturated fatty acid (C16:1) that acts as a lipokine, which is secreted and improves resistance to contract diabetes [33].

Recently, a comprehensive meta-analysis on longevity has been performed by Sebastiani et al. [34] by assembling a collection of 28,297 subjects from 7 different studies on long-living people and healthy ageing (see results of APOE above).

6.5   Whole Exome Sequencing (WES) and Whole Genome
    Sequencing (WGS)

The first sequencing technology has been developed by Frederick Sanger in the 1970s, and for this invention, he deserved his second Nobel Prize in chemistry in 1980. The Sanger sequencing took few decades to improve accuracy and speed up the process, since the arrival of next-generation sequencing (NGS), a new technology that allows to perform parallel sequencing with a high speed and low costs. In the last decade, costs for genome sequencing became more and more affordable (around 1000$ for an entire genome), and now, it is one of the most used techniques for unveiling the secrets of DNA. Different from genetic arrays, the target of sequencing analysis is the entire genome and not specific markers. There are two main approaches for the analysis of genetic information: WES and WGS. The WES approach consists in the sequencing of all the coding regions of the genome, which represent around 2% of the entire genome but contain the bigger amount of variability influencing gene function and disease. The WGS is comprehensive of all the genetic information in DNA, i.e., coding and non-coding regions. The latter are becoming more and more important in this kind of research because of the presence of regulatory and recombination sites.

The availability of this huge amount of data does not correspond to an easier and more efficient way to discover genetic variants associated with the phenotype. Especially for complex traits, to reach an adequate statistical power, NGS studies need very large populations of cases and controls because of the big number of genetic variants and because of the rare variants (see above).

Furthermore, data generated by NGS are very complex, and the amount of information for each sample is so large that it is difficult to manage without specialized bioinformatics personnel. Today, computer algorithms, web databases and artificial intelligence are very important tools for the analysis of the output data and for the prediction of the impact of genetic variants.

For the reasons mentioned above, very few association studies based on sequencing technology have been performed. The need of very large populations pushed researchers to use the NGS approach together with other phenotypic data.

However, around the world, centenarian populations took part in different studies such as the Genome of the Netherlands project, with 250 nuclear families [35, 36], or the SardiNIA project, with 1000 individuals [37–39], or the Wellderly study, involving 2000 individuals over 85 years of age [40]. The whole genomes of these populations have been sequenced to identify different genetic variants and to build databank for inter-population analyses.

More in detail, the first whole genome sequencing study on the genetics of human longevity has been performed by Sebastiani et al. [41] and it consists in the sequencing of 2 supercentenarians: a male and a female aged more than 114 years old. Obviously, no statistics have been performed on a such small number of samples, but the study gave interesting information about the genome of the two individuals. The female subject did not show to be a carrier of many common variants that have been associated with longevity in previous GWAS: only 5 of 16. Moreover, her genome showed to have large areas of homozygosity, probably due to inbreeding events among her ancestors, and this indicated that she selected a non-common genetic way to slow ageing. On the contrary, the male centenarian showed a larger concordance with GWAS carrying 11 of 16 common age-related variants. The second WGS study was carried out by Ye et al. [42] and it consisted in the genome sequencing of a pair of twins. This unique study did not identify an association with a particular variant, but highlighted the negative role of somatic variant accumulation during ageing and the ability of the analysed twins in limiting this accumulation. In 2014, Gierman et al. [43] sequenced 17 samples, but the study suffered the lack of a proper control population. In 2016, Erikson et al. [40] performed the WGS of the above-mentioned Wellderly population, which were mostly elderly individuals in good health with an age range of 84.2 ± 9.3 years. In summary, the sequencing approach, which is currently developing and improving, found very few and not always unitary results, with the association with longevity of a variant of Teashirt Zinc Finger Homeobox 3 (TSHZ3) gene in the first cohort analysed by Gierman group, and of Collagen Type XXV Alpha 1 Chain (COL25A1) gene in the Wellderly population. TSHZ3 is a transcription factor that regulates smooth muscle cell differentiation and the inhibition of caspase expression correlated with progression of Alzheimer's disease [44]. COL25A1 encodes a brain-specific membrane-associated collagen, and its malfunction can be responsible for Alzheimer amyloid plaque formation [45].

6.6   Conclusion

As shown in Fig. 6.1, scientists all over the world have been able to move from candidate gene studies based on the previous knowledge of the gene function to genome-wide fishing expeditions with the potentiality of finding genes whose function is unknown. The advancement has been possible thanks to the technological progress in terms of genome scan, data storage and statistical analysis. Despite the improvement made in controlling for stratification, the differences among studies in terms of environment of the population adopted and the criteria of inclusion have made difficult the replication of most of the findings. Still, elegant studies have made possible the identification of genetic signature that discriminate the morbidity and survival of individual carriers.

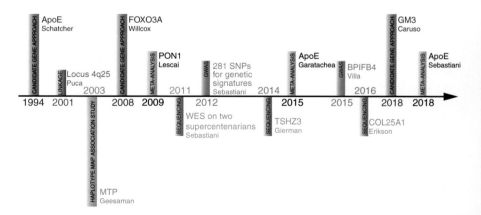

Fig. 6.1 The diagram shows the sequence of the major studies related to genetics of longevity depicting the used study design, the findings and the first author of the relative publications

References

1. Puca AA, Spinelli C, Accardi G, Villa F, Caruso C. Centenarians as a model to discover genetic and epigenetic signatures of healthy ageing. Mech Ageing Dev. 2018;174:95–102. https://doi.org/10.1016/j.mad.2017.10.004.
2. Ferrario A, Villa F, Malovini A, Araniti F, Puca AA. The application of genetics approaches to the study of exceptional longevity in humans: potential and limitations. Immun Ageing. 2012;9(1):7. https://doi.org/10.1186/1742-4933-9-7.
3. Perls TT, Wilmoth J, Levenson R, Drinkwater M, Cohen M, Bogan H, et al. Life-long sustained mortality advantage of siblings of centenarians. Proc Natl Acad Sci U S A. 2002;99(12):8442–7. https://doi.org/10.1073/pnas.122587599.
4. Lescai F, Marchegiani F, Franceschi C. PON1 is a longevity gene: results of a meta-analysis. Ageing Res Rev. 2009;8(4):277–84. https://doi.org/10.1016/j.arr.2009.04.001.
5. Schachter F, Faure-Delanef L, Guenot F, Rouger H, Froguel P, Lesueur-Ginot L, et al. Genetic associations with human longevity at the APOE and ACE loci. Nat Genet. 1994;6(1):29–32. https://doi.org/10.1038/ng0194-29.
6. Willcox BJ, Donlon TA, He Q, Chen R, Grove JS, Yano K, et al. FOXO3A genotype is strongly associated with human longevity. Proc Natl Acad Sci U S A. 2008;105(37):13987–92. https://doi.org/10.1073/pnas.0801030105.
7. Mohler PJ, Healy JA, Xue H, Puca AA, Kline CF, Allingham RR, et al. Ankyrin-B syndrome: enhanced cardiac function balanced by risk of cardiac death and premature senescence. PLoS One. 2007;2(10):e1051. https://doi.org/10.1371/journal.pone.0001051.
8. Malovini A, Illario M, Iaccarino G, Villa F, Ferrario A, Roncarati R, et al. Association study on long-living individuals from southern Italy identifies rs10491334 in the CAMKIV gene that regulates survival proteins. Rejuvenation Res. 2011;14(3):283–91. https://doi.org/10.1089/rej.2010.1114.
9. Anselmi CV, Malovini A, Roncarati R, Novelli V, Villa F, Condorelli G, et al. Association of the FOXO3A locus with extreme longevity in a southern Italian centenarian study. Rejuvenation Res. 2009;12(2):95–104. https://doi.org/10.1089/rej.2008.0827.
10. Garatachea N, Marin PJ, Santos-Lozano A, Sanchis-Gomar F, Emanuele E, Lucia A. The ApoE gene is related with exceptional longevity: a systematic review and meta-analysis. Rejuvenation Res. 2015;18(1):3–13. https://doi.org/10.1089/rej.2014.1605.
11. Di Bona D, Accardi G, Virruso C, Candore G, Caruso C. Association between genetic variations in the insulin/insulin-like growth factor (Igf-1) signaling pathway and longevity: a systematic review and meta-analysis. Curr Vasc Pharmacol. 2014;12(5):674–81.
12. Kops GJ, Dansen TB, Polderman PE, Saarloos I, Wirtz KW, Coffer PJ, et al. Forkhead transcription factor FOXO3a protects quiescent cells from oxidative stress. Nature. 2002;419(6904):316–21. https://doi.org/10.1038/nature01036.
13. Puca AA, Ferrario A, Maciag A, Accardi G, Aiello A, Gambino CM, et al. Association of immunoglobulin GM allotypes with longevity in long-living individuals from southern Italy. Immun Ageing. 2018;15:26. https://doi.org/10.1186/s12979-018-0134-7.
14. Geesaman BJ, Benson E, Brewster SJ, Kunkel LM, Blanche H, Thomas G, et al. Haplotype-based identification of a microsomal transfer protein marker associated with the human lifespan. Proc Natl Acad Sci U S A. 2003;100(24):14115–20. https://doi.org/10.1073/pnas.1936249100.
15. Puca AA, Daly MJ, Brewster SJ, Matise TC, Barrett J, Shea-Drinkwater M, et al. A genome-wide scan for linkage to human exceptional longevity identifies a locus on chromosome 4. Proc Natl Acad Sci U S A. 2001;98(18):10505–8. https://doi.org/10.1073/pnas.181337598.
16. Huffman DM, Deelen J, Ye K, Bergman A, Slagboom EP, Barzilai N, et al. Distinguishing between longevity and buffered-deleterious genotypes for exceptional human longevity: the case of the MTP gene. J Gerontol A Biol Sci Med Sci. 2012;67(11):1153–60. https://doi.org/10.1093/gerona/gls103.

17. Greenwood TA, Rana BK, Schork NJ. Human haplotype block sizes are negatively corre- lated with recombination rates. Genome Res. 2004;14(7):1358–61. https://doi.org/10.1101/ gr.1540404.
18. Gabriel SB, Schaffner SF, Nguyen H, Moore JM, Roy J, Blumenstiel B, et al. The struc- ture of haplotype blocks in the human genome. Science. 2002;296(5576):2225–9. https://doi. org/10.1126/science.1069424.
19. Sebastiani P, Solovieff N, Dewan AT, Walsh KM, Puca A, Hartley SW, et al. Genetic signatures of exceptional longevity in humans. PLoS One. 2012;7(1):e29848. https://doi.org/10.1371/ journal.pone.0029848.
20. Conneely KN, Capell BC, Erdos MR, Sebastiani P, Solovieff N, Swift AJ, et al. Human longev- ity and common variations in the LMNA gene: a meta-analysis. Aging Cell. 2012;11(3):475– 81. https://doi.org/10.1111/j.1474-9726.2012.00808.x.
21. Castro E, Edland SD, Lee L, Ogburn CE, Deeb SS, Brown G, et al. Polymorphisms at the Werner locus: II. 1074Leu/Phe, 1367Cys/Arg, longevity, and atherosclerosis. Am J Med Genet. 2000;95(4):374–80.
22. Gentschew L, Flachsbart F, Kleindorp R, Badarinarayan N, Schreiber S, Nebel A. Polymorphisms in the superoxidase dismutase genes reveal no association with human longevity in Germans: a case-control association study. Biogerontology. 2013;14(6):719–27. https://doi.org/10.1007/s10522-013-9470-3.
23. Hitt R, Young-Xu Y, Silver M, Perls T. Centenarians: the older you get, the healthier you have been. Lancet. 1999;354(9179):652. https://doi.org/10.1016/S0140-6736(99)01987-X.
24. Villa F, Carrizzo A, Spinelli CC, Ferrario A, Malovini A, Maciag A, et al. Genetic analysis reveals a longevity-associated protein modulating endothelial function and angiogenesis. Circ Res. 2015;117(4):333–45. https://doi.org/10.1161/CIRCRESAHA.117.305875.
25. Villa F, Carrizzo A, Ferrario A, Maciag A, Cattaneo M, Spinelli CC, et al. A model of evolu- tionary selection: the cardiovascular protective function of the longevity associated variant of BPIFB4. Int J Mol Sci. 2018;19(10) https://doi.org/10.3390/ijms19103229.
26. Spinelli CC, Carrizzo A, Ferrario A, Villa F, Damato A, Ambrosio M, et al. LAV-BPIFB4 iso- form modulates eNOS signalling through Ca2+/PKC-alpha-dependent mechanism. Cardiovasc Res. 2017;113(7):795–804. https://doi.org/10.1093/cvr/cvx072.
27. Puca AA, Carrizzo A, Villa F, Ferrario A, Casaburo M, Maciag A, et al. Vascular ageing: the role of oxidative stress. Int J Biochem Cell Biol. 2013;45(3):556–9. https://doi.org/10.1016/j. biocel.2012.12.024.
28. Villa F, Malovini A, Carrizzo A, Spinelli CC, Ferrario A, Maciag A, et al. Serum BPIFB4 levels classify health status in long-living individuals. Immun Ageing. 2015;12:27. https://doi. org/10.1186/s12979-015-0054-8.
29. Spinetti G, Sangalli E, Specchia C, Villa F, Spinelli C, Pipolo R, et al. The expression of the BPIFB4 and CXCR4 associates with sustained health in long-living individuals from Cilento- Italy. Aging. 2017;9(2):370–80. https://doi.org/10.18632/aging.101159.
30. Vecchione C, Villa F, Carrizzo A, Spinelli CC, Damato A, Ambrosio M, et al. A rare genetic variant of BPIFB4 predisposes to high blood pressure via impairment of nitric oxide signaling. Sci Rep. 2017;7(1):9706. https://doi.org/10.1038/s41598-017-10341-x.
31. Sebastiani P, Gurinovich A, Bae H, Andersen S, Malovini A, Atzmon G, et al. Four genome- wide association studies identify new extreme longevity variants. J Gerontol A Biol Sci Med Sci. 2017;72(11):1453–64. https://doi.org/10.1093/gerona/glx027.
32. Puca AA, Andrew P, Novelli V, Anselmi CV, Somalvico F, Cirillo NA, et al. Fatty acid profile of erythrocyte membranes as possible biomarker of longevity. Rejuvenation Res. 2008;11(1):63– 72. https://doi.org/10.1089/rej.2007.0566.
33. Frigolet ME, Gutierrez-Aguilar R. The role of the novel lipokine palmitoleic acid in health and disease. Adv Nutr. 2017;8(1):173S–81S. https://doi.org/10.3945/an.115.011130.
34. Sebastiani P, Gurinovich A, Nygaard M, Sasaki T, Sweigart B, Bae H, et al. APOE alleles and extreme human longevity. J Gerontol A Biol Sci Med Sci. 2018;74:44. https://doi.org/10.1093/ gerona/gly174.

35. Gorbunova V, Seluanov A, Mao Z, Hine C. Changes in DNA repair during aging. Nucleic Acids Res. 2007;35(22):7466–74. https://doi.org/10.1093/nar/gkm756.
36. Seluanov A, Mittelman D, Pereira-Smith OM, Wilson JH, Gorbunova V. DNA end joining becomes less efficient and more error-prone during cellular senescence. Proc Natl Acad Sci U S A. 2004;101(20):7624–9. https://doi.org/10.1073/pnas.0400726101.
37. Pilia G, Chen WM, Scuteri A, Orru M, Albai G, Dei M, et al. Heritability of cardiovascular and personality traits in 6,148 Sardinians. PLoS Genet. 2006;2(8):e132. https://doi.org/10.1371/journal.pgen.0020132.
38. Scuteri A, Sanna S, Chen WM, Uda M, Albai G, Strait J, et al. Genome-wide association scan shows genetic variants in the FTO gene are associated with obesity-related traits. PLoS Genet. 2007;3(7):e115. https://doi.org/10.1371/journal.pgen.0030115.
39. Chen WM, Abecasis GR. Family-based association tests for genomewide association scans. Am J Hum Genet. 2007;81(5):913–26. https://doi.org/10.1086/521580.
40. Erikson GA, Bodian DL, Rueda M, Molparia B, Scott ER, Scott-Van Zeeland AA, et al. Whole-genome sequencing of a healthy aging cohort. Cell. 2016;165(4):1002–11. https://doi.org/10.1016/j.cell.2016.03.022.
41. Sebastiani P, Riva A, Montano M, Pham P, Torkamani A, Scherba E, et al. Whole genome sequences of a male and female supercentenarian, ages greater than 114 years. Front Genet. 2011;2:90. https://doi.org/10.3389/fgene.2011.00090.
42. Ye K, Beekman M, Lameijer EW, Zhang Y, Moed MH, van den Akker EB, et al. Aging as accelerated accumulation of somatic variants: whole-genome sequencing of centenarian and middle-aged monozygotic twin pairs. Twin Res Hum Genet. 2013;16(6):1026–32. https://doi.org/10.1017/thg.2013.73.
43. Gierman HJ, Fortney K, Roach JC, Coles NS, Li H, Glusman G, et al. Whole-genome sequencing of the world's oldest people. PLoS One. 2014;9(11):e112430. https://doi.org/10.1371/journal.pone.0112430.
44. Kajiwara Y, Akram A, Katsel P, Haroutunian V, Schmeidler J, Beecham G, et al. FE65 binds Teashirt, inhibiting expression of the primate-specific caspase-4. PLoS One. 2009;4(4):e5071. https://doi.org/10.1371/journal.pone.0005071.
45. Kakuyama H, Soderberg L, Horigome K, Winblad B, Dahlqvist C, Naslund J, et al. CLAC binds to aggregated Abeta and Abeta fragments, and attenuates fibril elongation. Biochemistry. 2005;44(47):15602–9. https://doi.org/10.1021/bi051263e.

7

Dina Bellizzi, Francesco Guarasci, Francesca Iannone, Giuseppe Passarino, and Giuseppina Rose

7.1 DNA Methylation

DNA methylation represents the most prevalent epigenetic modification in all kingdoms of life and comprises a covalent transfer of methyl group to the aromatic ring of the DNA nitrogenous base [1–3].

C5-methylcytosine (5-mC) is the canonical methylated base in eukaryotes, N6-methyladenosine (m6A) is the dominant modification in bacteria, meanwhile N4-methylcytosine (4-mC) is very common in bacteria but absent in mammals. Albeit it has also been hypothesized, the presence of 6mA in eukaryotic genomes and its minimal levels are detectable only by highly sensitive methods [4–7].

In vertebrates, methylation mostly occurs at the cytosines followed by guanine residues (CpG methylation), although recent data report the presence of methylation in embryonic stem cells and neurons at sites other than CpGs (non-CpG methylation), mainly in CpA context, likely regulating cell type-specific functions [8–10].

Notably, methylated CpGs are predominantly located into intergenic and intronic CpG-poor regions and repetitive sequences, such as interspersed and tandem repeats, most of which are derived from transposable elements. Unmethylated CpG dinucleotides are, instead, concentrated in CpG-rich regions, termed CpG islands (CGIs),which are, on average, 1000 base pairs long and show an elevated G + C base composition and little CpG depletion [11–14]. Approximately, CpG islands have been demonstrated to be associated with 70% of the annotated gene promoters, including all housekeeping genes, a number of tissue-specific genes and developmental regulator genes [14–20].

Dina Bellizzi, Francesco Guarasci, Francesca Iannone, Giuseppe Passarino, and Giuseppina Rose, who are listed in alphabetical order, equally contributed to this work.

D. Bellizzi · F. Guarasci · F. Iannone · G. Passarino (✉) · G. Rose
Department of Biology, Ecology and Earth Science, University of Calabria, Rende, Italy
e-mail: dina.bellizzi@unical.it; francesco.guarasci@unical.it; giuseppe.passarino@unical.it; pina.rose@unical.it

Methylation patterns come from the activity of enzymes belonging to the family of DNA methyltransferases (DNMTs), which transfer a methyl group from *S*-adenosyl-L-methionine (SAM) to deoxycytosine. In particular, DNMT1 is involved in the maintenance of DNA methylation during cell division by acting on hemi-methylated CpG sequences, and DNMT3a and 3b are both responsible of the de novo establishment of DNA methylation [21–23].

The dynamic regulation of the genome is determined by the balance between the events of DNA methylation and demethylation. The latter process includes both the loss of 5mC during the replication (passive demethylation), induced by downregulation of DNMT enzymes, inhibition of their activity or decreased levels of SAM, and the active removal of 5-mC (active demethylation) resulting in the formation of 5-hydroxymethylcytosine (5-hmC), considered to date the sixth base of DNA and a novel epigenetic mark [24–27]. Recently, Penn et al. discovered for the first time in mouse and frog brain that the presence of 5-hmC has been reported in different tissues and cells and considered as an intermediate of the oxidation of 5-mC by ten-eleven translocation (TET)-family of methyl-cytosine dioxygenases [28–35].

The earliest observations of the DNA methylation function date back to transfection experiments and microinjections of methylated sequences, demonstrating that it induces gene silencing and that, in cultured cell lines, silent genes can be activated following the treatment with the demethylating agent 5-azacytidine [36–42].

Genome-wide studies of methylome have highlighted that methylation patterns are cell-type specific and that their effects are influenced by the position of methylated cytosines: if located adjacent to transcription factor binding sites, they block initiation, through either recruiting specific factors acting as gene expression repressors or inhibiting the binding of activators, meanwhile in body gene, they may either stimulate transcription elongation, thus hypothesizing a their role on splicing, or impede the alternative promoters activation [20, 43–48]. Methylation in repeat regions such as centromeres is important for chromosomal stability, for example chromosome segregation at mitosis, and is also likely involved in the suppression of the expression of transposable elements and thus to have a role in genome stability. Recently, the role of methylation in altering the activities of enhancers, insulators and other regulatory regions has been described. CpG island methylation of the transcription start sites is associated with long-term silencing, for example chromosome X inactivation, imprinting, genes expressed predominantly in germ cells and some tissue-specific genes [14, 49, 50].

Moreover, a more complex epigenetic landscape is emerging, as demonstrated by the role played by the mitochondrial genome (mtDNA) in regulating intracellular DNA methylation as well as by the evidence reporting that, similar to nuclear genome, mtDNA is also subject to CpG, non-CpG methylation and hydroxymethylation, all events strongly influenced by environmental factors, nutrition and drugs [51–55].

7.2   Histone Modifications

In eukaryotes, DNA molecule is structured in chromatin, a higher-order nucleopro-tein complex where the basic unit is the nucleosome containing 147 base pairs of DNA wrapped around an octamer of two of each core histone protein (H2A, H2B, H3 and H4). Each nucleosome is linked to the next by small segments of linker DNA and the binding of histone H1 to both this linker and each nucleosome stabi-lizes the polynucleosome fibre [50, 56, 57].

The structure of the chromatin may be altered in its constituents, and this altera-tion is fundamental in the regulation of gene expression since it determines the accessibility and the recruitment of regulatory factors [50, 58–60].

Histones undergo several epigenetic post-translational modifications, also referred to as histone marks, including lysine acetylation, methylation, ubiquitina-tion and sumoylation, arginine methylation, serine and threonine phosphorylation, glutamate ADP-ribosylation and proline isomerization [50, 60–62]. The modifica-tion status of histone is modulated by enzymes with specific catalytic activities: histone modifying (writers), which catalyse the transfer of chemical groups to amino acid residues on histones, recognition enzymes (readers), which interpret modification signals, and de-modifying enzymes (erases), which remove the modi-fications [63–65].

Histone modifications not only have a functional role by merely being there but, by recruiting remodelling ATP-dependent enzymes, regulate chromatin compaction and nucleosome dynamics, thereby controlling various DNA-templated processes, such as gene transcription, DNA replication, DNA recombination, and DNA repair which contribute to basic cellular functions, including cell cycle, cell growth and apoptosis [65–68]. Specific histone modification patterns correlate with chromatin elements, and maps of the modifications along a chromosome can be predictive of both eu- and hetero-chromatin and transcriptional activity. For example, trimethyl-ated histone 3 on lysine 9 (H3K9) modification appears to localize largely to hetero-chromatin, whereas acetylated H3K9 and histone 3 on lysine 16 (H3K16) are often found within euchromatic sites [62, 69–72].

The most well-characterized histone modifications are acetylation and methyla-tion. Histone acetylation is considered a central switch that allows interconversion between permissive and repressive chromatin domains in terms of transcriptional activity through the addition and removal of acetyl groups from acetyl-CoA to lysine residues by histone acetyltransferases (HATs), including GNAT (Gcn5-related N-acetyltransferase) superfamily, MYST (named after its founding members monocytic leukaemia zinc-finger [MOZ], Ybf2/Sas3, Sas2 and Tip60), p300/CBP (p300/CREB-binding protein) and TFIII C (transcription factor III C) families and histone deacetylases (HDACs), such as the Sirtuin family [73–78].

Histone methylation comprises the addition of methyl groups on lysine, cata-lysed by histone methyltransferases (HMTs), or on arginine residues, catalysed by protein arginine methyltransferases (PRMTs), in one of three different forms:

mono-, di- or trimethyl for lysines and mono- or dimethyl for arginines [79–82]. The methylation of both residues has been described to be associated with gene activation, for example histone H3 lysine 4 (H3K4), histone H3 lysine 36 (H3K36), histone H3 lysine 79 (H3K79), histone 4 arginine 3 (H4R3), histone 4 arginine 17 (H3R17) or repression, including histone H3 lysine 9 (H3K9), histone H3 lysine 27 (H3K27), histone H4 lysine 20 (H4K20), and histone 3 arginine 8 (H3R8) [66, 79, 83]. Histone demethylases have also been identified, such as LSD1 demethylates H3K4 and H3K9 thus repressing the transcription or JHDM family members demethylate H3K9 and H3K36 [68, 84–86].

Recent findings showed that a cross-talk between DNA methylation and histone modification occurs, mostly mediated by a group of proteins with methyl DNA binding activity, including methyl CpG binding protein 2 (MeCP2), methyl-CpG binding domain protein 1 and ZBTB 33 (zinc finger and BTB domain containing protein 33) which localize to DNA-methylated promoters and recruit a protein complex that contains histone deacetylases (HDACs) and histone methyltransferases. For example, the lysine methylation that contributes to reversible regulation of gene expression in several cellular processes often contrasts with stable gene inactivation by DNA methylation [87–91].

7.3   Non-coding RNAs

The last decade has been characterized by a growing interest for non-coding RNAs because of their potential role in several biological processes. There are several types of non-coding RNAs such as small nuclear RNA (snRNA), long non-coding RNA (lncRNA), silencing RNA (siRNA) and microRNA (miRNA) [92].

The latter, in particular, have been extensively studied for their role in regulating the expression of a wide array of genes. miRNAs are small non-coding single-stranded RNAs, approximately 21–25 nucleotides long, highly conserved. They may regulate gene expression binding to their complementary messenger RNAs (mRNAs) and preventing the production of the corresponding protein.

Two independent groups of researchers, Lee et al. and Wightman et al., first discovered miRNAs in *Caenorhabditis elegans* in 1993, resulting in significant breakthroughs in genetics research. They observed that lin-4 mRNA was not translated into a protein, but it gave rise to two small RNAs, 21 and 61 nucleotides, respectively, long. The 61-nucleotide RNA was the precursor for the shorter one, forming a stem-loop structure. Later, they found that the smaller RNA had antisense complementarity to 3′ UTR of lin-14 mRNA, and this binding led to lin-14 downregulated protein, essential for the progression from the first larval stage (L1) to the second (L2) in *C. elegans* [93]. Since then, researchers thought that this kind of regulation of gene expression was a characterizing phenomenon of *C. elegans*, and they continued detecting other small RNAs, such as let-7, necessary for the development from a later stage to adult in *C. elegans*. Some small RNAs have been found to be conserved across species, and therefore, new miRNAs were discovered in different organisms.

Let-7, first revealed in *C. elegans*, was identified as the first human miRNA by an alignment research. In humans, the let-7 family includes nine members, encoded by 12 different genomic loci, whose functionality has not been fully understood yet [94].

The rapid growth of the miRNA field led to the creation of miRNA registries to collect all miRNA sequences, annotation data and predictive information. They contain entries of precursor miRNAs and mature miRNA products in different species. The experimentally confirmed miRNAs are numbered, preceded by the prefix 'miR' and by three letters indicating the organism (e.g. hsa for *Homo sapiens*).

miRNA biogenesis is an intricate process, characterized by several steps, that involves several protein complexes which regulate the formation of mature miRNA. miRNAs are initially transcribed by RNA polymerase II as several hundred nucleotide long primary transcripts (pri-miRNAs) with a 5' guanosine cap and a 3' polyadenylated tail. The pri-miRNA is then processed into a pre-miRNA (~70- to 120-nucleotide long) by a multiprotein complex containing a ~160-kDa nuclear RNase III enzyme, called Drosha. The pre-miRNA is then exported into the cytoplasm by exportin 5 where it is finally processed into a mature duplex (~18- to 23-nucleotide long) by another RNase III enzyme, Dicer-1. The strands of the duplex are then separated: the guide strand results in an unstable base pairing at the 5' end, and it associates with Argonaute (AGO) proteins, forming a ribonucleaoprotein complex called RNA-induced silencing complex (RISC). On the contrary, the second strand, called the passenger strand, which results in a perfect base pairing, is usually degraded [95]. The guide strand directs the RISC complex to the target mRNA through sequence complementarity and causes its translational repression. However, in some cases, miRNA biogenesis can be Drosha-complex independent such as for pre-miRNA-like hairpins and some small nucleolar RNAs (snoRNAs) [96].

miRNAs mediate post-transcriptional control of the expression of their target genes by binding to their complementary mRNA. The recognition of the target site is mostly determined by the 'seed region' (nucleotides 2–8) at the 5' end of the mature miRNA [97]. The main site for miRNA binding is the 3' UTR of an mRNA: mutations in this region render the corresponding mRNA insensitive to miRNA regulation. Recent evidence shows that miRNAs can also bind the target sequences in 5' UTR and in the open reading frame [98]. Each miRNA can target a large number of mRNAs, and one mRNA can be regulated by several miRNAs [97].

Mature miRNAs execute the post-transcriptional repression of their target mRNAs in two ways: reducing the translational efficiency or decreasing the mRNA levels. High miRNA-mRNA complementarity results in mRNA degradation, while less complementarity leads to translational inhibition [99]. Moreover, the binding of miRNAs seems to lead a faster deadenylation of mRNAs, accelerating their degradation [100].

In addition to intracellular miRNAs that regulate several functions inside the cell, circulating miRNAs have been detected [101]. These extracellular miRNAs are present in different biofluids including blood, plasma, serum, saliva, urine and pleural effusions. In contrast to cellular miRNAs, circulating miRNAs are more stable and contribute to cell–cell communication, regulating important developmental mechanisms. One of the responsible pathways for the different expression of circulating miRNAs seems to be the transforming growth factor-beta (TGF-β) signalling that acts in modulating miRNA maturation [102]. miRNAs can be transported from cell to cell and surprisingly from one generation to the next because they can enter the germ line. The first discovered mobile miRNAs were from neurons to the germ line in *C. elegans* [103]. The exact mechanism of release of cellular miRNAs to extracellular medium is not deeply known. They can be distributed as free molecules or bound to carriers from donor cells to recipient cells. The carriers of miRNAs are high-density lipoproteins (HDL), low-density lipoproteins (LDL), ribonucleoproteins and extracellular vesicles (EVs) [101]. Extracellular vesicles (EVs) are small membrane vesicles, enriched for proteins, cytokines, mRNAs, lipids, miRNAs and long non-coding RNAs [104]. Many of these circulating miRNAs seem to be stored exclusively in the EVs and selectively released. Moreover, the expression profile of circulating miRNAs from different biofluids is specific in relation to different pathophysiological conditions [105].

7.4   Epigenetic Modifications in Ageing

7.4.1   DNA Methylation and Ageing

Ageing is a slow and gradual decline process of functional abilities that makes individuals more susceptible to environmental phenomena and diseases and leads to a reduction in the probability of survival and finally to death [106–108].

Ageing affects all living organisms but lifespan is characteristic of each species. Moreover, among the various populations and within them, there is a considerable variability with regard to the way and the quality of ageing. This heterogeneity has largely been described as resulting from a complex interaction among genetics, environmental and stochastic factors, and, more recently, epigenetic alterations have been included [109, 110]. These alterations, by regulating gene expression, influence not only most of hallmarks of ageing (genomic instability, telomere attrition, loss of proteostasis, deregulated nutrient sensing, mitochondrial dysfunction, cellular senescence, stem cell exhaustion, altered intercellular communication) but, at the same time, themselves because they are subjected to dynamic changes during lifetime, a phenomenon described as epigenetic drift [111–113]. During early embryogenesis, genomic DNA undergoes reprogramming processes including genome-wide demethylation and de novo methylation leading to the re-establishment of DNA methylation patterns in the progeny that will be maintained in the somatic cells throughout the lifespan. After birth, although global DNA methylation patterns are quite stable, stochastic and environmental stimuli (ROS, inflammation, diet) as well as the failure of the epigenetic machinery may induce random changes at certain loci, leading to a loss of phenotypic plasticity among individuals [114–116]. Indeed, during each cell division, aberrant DNA methylation patterns accumulate over time contributing to epigenetic drift and creating an epigenetic mosaicism that may allow for the selection of biological defects that may lead to cancer and other age-related diseases [117]. Support for these evidences comes mainly from studies carried out in mono- and di-zygotic twin pairs in which a gradual age-related divergence in epigenetic marks was observed in monozygotics [118–123]. DNA methylation drift comes from non-directional changes occurring during ageing and involves both hypermethylation and hypomethylation events. Recently, Slieker and colleagues identified several age-related Variably Methylated Positions (aVMPs) exhibiting high variability in their methylation status, which are associated with the expression of genes involved in DNA damage and apoptosis [124]. During ageing, epigenetic drift also deeply influences the function of aged stem cells by limiting their plasticity and their differentiation potential that ultimately results in the exhaustion of the stem cell pool and in the selective growth advantage in other stem cells, which leads to clonal expansion and local hyperproliferation [113, 125, 126].

Recently, several studies reported the presence of directional and non-stochastic changes occurring over time within clusters of consecutive CpG sites throughout the whole genome, referred as age-related Differentially Methylated Regions (a-DMRs) [113, 127, 128]. Hundreds of hyper- and hypo-methylated a-DMR have been identified in multiple tissues and replicated in independent samples. Literature data agree to consider them associated with biological mechanisms involved in ageing and longevity. Ashapkin et al. assume that most hyper-aDMRs represent epigenetic perturbations inherent to ageing per se, while hypo-aDMRs may be correlated to modifications associated with both ageing per se and age-dependent modifications in relative proportions of the blood cell subtypes [129, 130].

Candidate genetic loci undergoing profound epigenetic changes with age and in age-related diseases have been progressively characterized. Global genomic DNA hypomethylation is especially evident at repetitive sequences, to a greater extent at Alu and HERV-K sequences, contributing to the increase of genome instability as well as at specific promoter regions of some genes including ITGAL (Integrin alpha-L) and IL17RC (Interleukin 17 Receptor C) [131–135]. By whole-genome bisulfite sequencing (WGBS), Heyn et al. [136] compared the DNA methylation state of more than 90% of all CpGs present in the genome between newborn and nonagenarian/centenarian samples. A significant loss of methylated CpGs was found in the centenarian versus newborn DNAs. This was observed for all chromosomes and concerned all genomic regions such as promoters, exonic, intronic and intergenic regions. Most of these changes were focal and the aged genome was consequently less homogeneously methylated with respect to the newborn due to the age-dependent epigenetic drift [136].

Besides this extensive hypomethylation, the promoter regions of specific genes are subjected to a gradual increase of DNA methylation across lifespan (Table 7.1). In most of the cases, the observed hypermethylation was associated with the transcriptional silencing, suggesting that with increasing age, there is an epigenetic turning off of these genes.

Epigenome-wide association studies (EWAS) identified the so-called clock CpGs, namely a large set of CpG markers whose methylation status is measured in order to construct quantitative models effective in predicting the age of cells, tissues

Table 7.1 List of genes showing age-related DNA methylation changes in CpG islands located within their promoter regions

Function	Gene symbol	References
Angiogenesis	VASH1	Reynolds et al. [218]
Antigen processing and presentation	DPB1, DRB1, LAG3, TAP2, PSMB9, PSMB8, HLA-E, HLA-F, HLA-B, MICB, SLC11A1, HLA-DPA1, TAPBP, HLA-DMB	Reynolds et al. [218]
Cell adhesion	LAMB1, PCDHA1,2,3,4, PODXL, PCDH9, SORBS2	McClay et al. [219]
Development and growth	c-Fos, FGF8, FIGN, IGF2, HOXB5, B6, B7, B8, MEIS1, MYOD1, NKX2-2, TIAL1, UBE2E3	Choi et al. [220] Issa et al. [221] Ahuja et al. [222] Christensen et al. [223] McClay et al. [219] Vidal et al. [224]
Genome stability and repair	MGMT, MLH1, OGG, hTERT, RAD50	Nakagawa et al. [225] Matsubayashi et al. [226] Silva et al. [227] Christensen et al. [223] Madrigano et al. [228]
Ion channel	GRIA2, KCNJ8, RYR2	McClay et al. [219]
Metabolism	AGPAT2, ATP13A4, COX7A1, CRAT, ECRG4, ELOVL2, EPHX2, GAD2, LEP, MGC3207, MGEA5, SLC38A4, SLC22A18, SNTG1, STAT5A	Rönn et al. [229] Bell et al. [129] Madrigano et al. [228] McClay et al. [219] Gentilini et al.
Immune response	CD4, INFG, TNFα, NOD2, PTMS	Madrigano et al. [228] McClay et al. [219]
Signal transduction	ARL4A, DLC1, GPR128, GRIA2, LAG3, MYO3A, PRR5L, PTPRT, TFG, TRAF6, TRHDE	Bell et al. [129] McClay et al. [219]
Stress response	HSPA2	McClay et al. [219]
Transcription factors	ARID5B, BICC1, ESR1, FOXP1, HIPK2, LHX5, MLF2, NFIA, NOD2, POU4F3, RARB, TBX4, TBX20, TRPS1, WT1, ZBTB1, ZEB2, ZNF827	Gaudet et al. [230] Christensen et al. [223] Bell et al. [129] Reynolds et al. [218] McClay et al. [219]
Tumour suppression	APC, CASP8, CHD1, GSTP1, HIC1, LOX, LSAMP, N33, P16INK4A, RASSF1, RUNX3, SOCS1, TIG1, DAPK1, hMLH1, p16	Ahuja et al. [222] Fujii et al. [231] Cody et al. [232] Dammann et al. [233] Virmani et al. [234] Waki et al. [235] Sutherland et al. [236] So et al. [237] Nishida et al. [238] Yuan et al. [239] Christensen et al. [223] McClay et al. [219]

or organs, referred as epigenetic age or DNAm age. DNAm age not only reflects the chronological but also the biological age, and thus, these biomarkers would, on the one hand, facilitate the differentiation of individuals who are of the same chronological age yet have variant ageing rates, and on the other define a panel of measurements for healthy ageing and, even further, predict life span [137–140]. Starting from Bocklandt et al., which described the first age estimator model by using DNA samples from saliva, a series of epigenetic clocks were developed by analysing DNA methylation marks in single and multiple tissues [141]. Currently, Hannum and Horvath clocks represent the most robust recognized models, both showing a high age correlation ($R > 0.9$) and low mean error of the age prediction (4.9 and 3.6 years, respectively). The first model was developed only in blood, while the second was designed matching data from 51 healthy tissues, such as blood, cerebellum, occipital cortex, buccal, colon, adipose, liver, lung, saliva and cell types, including CD4 T and immortalized B cells, which is compatible with different technological platforms and used in a wide range of studies [142, 143]. Both the models are also able to predict all-cause mortality independent of several risk factors including smoking, alcohol use, education, body mass index and comorbidities [144–146]. More recently, DNAm age biomarkers which also consider clinical measures of physiological dysregulations have been developed. One of these referred as DNAmPhenoAge (phenotypic age estimator), constructed by generating a weighted average of 10 clinical characteristics, such as albumin, creatinine, glucose and C-reactive proteins, and then analysed by regression analysis against DNA methylation levels in blood, was proved to be effective in predicting mortality, health span and disease risk and in various measures of comorbidity [147].

Environmental factors including chemicals, pollutants, diet, drugs, infectious, trauma and psycho-social and socio-economic status have been associated with DNA methylation changes with ageing [148, 149]. A summary of the above factors and their effects are reviewed in Table 7.2. Interestingly, it has emerged that the early intrauterine period of life is particularly sensitive to their exposure since DNA methylation patterns at DMRs are established before gastrulation. For example, energy-rich, protein-deficient, micronutrient-deficient and/or methyl donor-rich diets during pregnancy induce modifications of methylation profile in mothers which, in turn, can be transmitted to next generation, thus regulating long-term metabolic processes in offspring which contribute to age phenotypes and age-related diseases [150–152]. An emblematic example of the trans-generational relationship between food and epigenetic modifications is represented by The Dutch Hunger winter (1944–1945) family studies in which adult health outcomes in relation to exposure to famine prior to conception or at specific periods of gestation was analysed. It emerged that prenatal exposure to the famine is associated with increased prevalence of overweight, hypertension and coronary heart disease, meanwhile maternal famine exposure around the time of conception has been related to prevalence of major affective disorders, antisocial personality disorders, schizophrenia, decreased intracranial volume and congenital abnormalities of the central nervous system [153, 154].

Table 7.2 DNA methylation aberrations associated with environmental factors

Environmental factor		DNA methylation change	Target tissue	Reference
Chemicals and pollutants	Arsenic	↑ Global DNA methylation	Human blood	Pilsner et al. [240], Majumdar et al. [241]
		Differentially methylated DNA loci (744) and regions (15)	Exposure across human generations	Guo et al. [242]
		579 differentially methylated CpG sites	In utero exposed newborn blood cord	Kaushal et al. [243]
	Cadmium	1945 differentially methylated DNA loci, 15 of which involving imprinting control regions	In utero exposed blood cord	Cowley et al. [244]
		↑ ApoE	TRL 1215 cells	Hirao-Suzuki et al. [245]
		↓ MGMT; ↓ MT2A and DNMT3B in women; ↑ LINE-1 in men	Urine and blood human samples	Virani et al. [246]
	Chromium	↓ Global DNA methylation	Human blood	Wang et al. [247]
		↓ MT-TF; ↓ MT-RNR1	Human blood	Yang et al. [248]
		↓ Global DNA methylation; ↑ TP16	Human lymphoblastoid and lung cell lines	Lou et al. [249]
	Mercury	↑ Ratio of %-5mC to %-5hmC	Newborn blood in utero exposed	Cardenas et al. [250]
		↓ PON1	In utero exposed blood cord	Cardenas et al. [251]
		↓ SEPP1	Human buccal mucosa	Goodrich et al. [252]
	Lead	↓ 5mC levels at CpG sites; ↑ 5hmC levels at CpG sites	Human bronchial epithelial cells	Zhang et al. [253]
		↓ PRMT5	A549 and MCF-7 cells	Ghosh et al. [254]
		↑ XRCC1, ↑ hOGG-1, ↑ BRCA1, ↑ XPD	TK6 cells	Liu et al. [255]
	Air pollution	↑ Bdnf	Offspring brain of early life exposed rats	Cheong et al. [256]
		13 CpG sites and 69 DMRs	Human blood	Mostafavi et al. [257]
		↑ Global DNA methylation	Placenta	Maghbooli et al. [258]
		↑ Circadian pathway genes	Placenta	Nawrot et al. [259]
	Polycyclic aromatic hydrocarbons	↑ DNA methylation age	Human blood	Li et al. [260]
		↓ LINE	Human cord blood samples	Lee et al. [261]
		↑ CDH1; ↑ HIN1; ↑ RARβ; ↓ BRCA1	Breast tumour tissue	White et al. [262]
		↓ LINE	Human blood	White et al. [263]
	Smoke	11 DMRs	Human blood	Prince et al. [264]
		30 DMRs	In utero exposed cord blood	Witt et al. [265]
		↓ Global DNA methylation in lung; ↑ IFN-γ ; ↑ Thy-1	In utero and early life exposed rats	Cole et al. [266]

Table 7.2 (continued)

Environmental factor		DNA methylation change	Target tissue	Reference
Diet	Calorie restriction	Global DNA methylation remodelling	Mouse blood of different age	Sziráki et al. [267]
		↓ CpG islands; ↓ repetitive regions; ↑ 5′-UTR, exon and 3′-UTR	Aged rat kidney	Kim et al. [268]
		↓ Global DNA methylation; 477 DMRs	Mouse placenta	Chen et al. [269]
	Methyl-donor diet	DNA methylation changes in genes related to growth (IGF2), metabolism (RXRA) and appetite control (LEP)	Buccal cell DNA	Pauwels et al. [270]
		↑ Global DNA methylation and hydroxymethylation	Human blood	Pauwels et al. [271]
		431,312 DMRs	Human blood	Kok et al. [272]
	Bioactive com-pounds	↓ Tumour suppressor genes (p16INK4a, RARβ, MGMT, hMLH1)	Human oesophageal cancer KYSE 510 cells	Fang et al. [273]
		↑ 76 DMRs and ↓ 89 DMRs in pregnant women; ↑ 200 DMRs and ↓ 102 DMRs in post-partum mothers	Human blood	Anderson et al. [274]
		↑ Global DNA methylation; ↓ LINE-1	Colorectal cancer cell line Caco-2	Zappe et al. [275]
	High-fat diet	↑ Global DNA methylation;	Placenta and foetal liver	Ramaiyan and Talahalli [276]
		3966 DMRs, clustered in type 2 diabetes mellitus and adipocytokine signalling pathways	Rat liver	Moody et al. [277]
		↓ 12,494 DMRs; ↑ 6404 DMRs	Rat liver	Zhang et al. [278]

(continued)

Table 7.2 (continued)

Environmental factor		DNA methylation change	Target tissue	Reference
Drugs	Metham-phetamine	↑ Circadian clock genes	Rat brain	Nakatome et al. [279]
		↑ α-Synuclein	Rat brain	Jiang et al. [280]
		↑ SINE elements	Rat brain	Itzhak et al. [281]
		↓ GluA1; ↓ GluA2	Rat brain	Jayanthi et al. [282]
	Cocaine	↑ PPC1	Mouse	Anier et al. [283]
		↑ PP1Cβ	Mouse brain	Pol Bodetto et al. [284]
		↑ Global DNA methylation	Rat and mouse brain	Tian et al. [285]
		↑ Cdkl5	Rat brain	Carouge et al. [286]
		↓ FosB	Mouse brain	Ajonijebu et al. [287]
	Opioids	↑ OPRM-1; ↑ Sp1 transcription factor	Human blood, sperm	Ebrahimi et al. [288], Chorbov et al. [289]
		↑ ABCB1, ↑ CYP2D6; ↑ OPRM-1	In utero exposed human infants	McLaughlin et al [290]
		↑ POMC	Human blood	Groh et al. [291]
		↑ VEGF-A; ↑ NGF	Human blood	Groh et al. [292]
	Cannabis	DMR-associated genes involved in glutamatergic synaptic regulation	Rat brain	Watson et al. [293]
		↑ DRD2; ↑ NCAM1	Human blood	Gerra et al. [294]
	Alcohol	PDYN	Human brain	Taqi et al. [295]
		Gender-specific changes in MAOA	Human blood	Philibert et al. [296]
		↑ AVP; ↓ ANP	Human blood	Glahn et al. [297]
		↑ SNCA	Human blood	Foroud et al. [298]
		↓ 5hmC levels	Rat models for alcoholic liver disease (ALD)	Ji et al. [299]
		System-wide epigenetic changes in genes involved in cognition and attention-related processes	Human prenatal alcohol exposure	Frey et al. [300]
		59 differentially methylated CpG sites; ↓ global DNA methylation	Human blood	Brückmann et al. [301]
		↑ PHOX2 A	Human blood	Weng et al. [302]
		↑ NGF	Human blood	Heberlein et al. [303]
		↓ GDAP1	Human blood	Brückmann et al. [304]
		↓ DAT	Human blood	Jasiewicz et al. [305]

Table 7.2 (continued)

Environmental factor	DNA methylation change	Target tissue	Reference
Infections	↑ CADM1, ↑ MAL, ↑ DAPK1	Human cervical sample with intraepithelial neoplasia	Fiano et al. [306]
	DMRs in olfactory transduction, transport regulation, amino acid phosphorylation, phosphorus and programmed cell death genes	HBV-infected patients	Jin et al. [307]
	↑ Global DNA methylation, ↑ AURKB; ↑ AURKC; ↑ DNMT3B	CD4+ T cells infected with HIV-1	Nunes et al. [308]
	↑ Horvath's DNA methylation age	H. pylori-infected human sample	Gao et al. [309]
	↓ Global DNA methylation; ↓ ABCB1	Human blood sample from malaria subjects	Gupta et al. [310]
Trauma	17 DMRs	Human blood from post-traumatic stress disorder	Mehta et al. [311]
	↑ BDNF, ↑ NR3C1, ↑ SLC6A4, ↑ MAOB	Human blood from MZ twins with childhood trauma and depression	Peng et al. [312]
	↑ Hannum's DNA methylation age	Human blood from post-traumatic stress disorder	Wolf et al. [313]
	↓ DMRs in inflammatory genes and in immune and gene expression regulation-associated pathways	Combined bioinformatic analysis on osteoarthritis DNA data	Song et al. [314]
Psicosocial and socioeconomic factors	↑ Horvath's DNA methylation age	Human sample grouped according to range of social position	Hughes et al. [315]
	↑ SLC6A4	Human sample from low socio-economic status	Swartz et al. [316]
	↑ Global DNA methylation	Human sample from samples with asthma and low socio-economic status	Chan et al. [317]
	↑ Sat2; ↑ Alu	Adult human sample with low family income at birth	Tehranifar et al. [318]

Lastly, a significant number of reports evidence a correlation between mitochondrial DNA methylation with ageing. The first evidence dates to 1983, when a decrease of mtDNA methylation was observed in aged cultured fibroblasts [155]. More recently, although methylation levels of the mitochondrial D-loop region are not associated with ageing, high methylation levels (>10%) of one CpG site located within the MT-RNR1 gene were observed more frequent in old women with respect to youngers and appeared to be correlated with survival chances [51, 156].

Also, the non-canonical CpG methylation patterns, such as non-CpG and hydroxymethylation, are deregulated during ageing, potentially leading to downstream changes in transcription and cellular physiological functions. A global non-CpG methylation decrease with age has been described. 5-hmC content significantly decreases in some tissues, including blood and liver, and is negatively correlated with ageing, in association with low mRNA expression levels of TET1 and TET3 [157]. By contrast, mouse cerebellum and hippocampus show an increase of 5-hmC levels with ageing which can be prevented by caloric restriction [158, 159]. A decrease of mitochondrial DNA levels of 5-hmC during ageing was observed in frontal cortex but not in the cerebellum [160]. An increase in 5-hmC signals was observed in genes activated in old mice with respect to young ones demonstrating that 5-hmC is acquired in developmentally activated genes [158]. Furthermore, age-related non-overlapping 5-mC and 5-hmC pattern have been observed [161].

7.4.2   Histone Modifications and Ageing

Several studies demonstrated that histone modifications exert a pivotal role in influencing the expression of specific genes involved in longevity and in signalling pathways that modulate lifespan as well as in the phenotypic heterogeneity that characterizes ageing process. Indeed, similar to DNA methylation events, a loss of control in histone modifications has been observed with age [162, 163]. The epigenetic landscape of histone modifications appears even more complex if one considers the heterogeneity of the enzymes involved in these modifications (writers, readers and erasers) as well as all proteins, which form complexes with the above enzymes and are important for the recruitment of the complexes to specific genomic loci. It should not be overlooked, also, that an age-progressive loss of histone biosynthesis, with a consequent heterochromatin loss and a genome-wide decrease of nucleosome occupancy, is a conserved feature from yeast to humans, the global impact of which is to date poorly understood but certainly influences the levels of the modifications involving histone proteins [164–166].

Changes in the levels of histone mono-, di- and tri-methylation and acethylation as well as in the distribution of many histone methylation and acetylation marks and in the activity of all the histone-modifying enzymes have been identified in cellular and organismal models of ageing differently impacting lifespan, with a general effect that is frequently specie- and genomic context-specific [72, 166].

Specific mention should be made of important well-known family of nicotine adenine dinucleotide (NAD+)-dependent enzymes, referred to as sirtuins, initially described as transcription-silencing histone deacetylases in yeast and then identified in mammals, in which seven sirtuins (SIRT1-7) exhibit differentially subcellular localization, substrate affinity and activity. Indeed, next to their role as deacetylases of lysine residues, some of mammal sirtuins also have deacylase and O-ADP-rybosylase activity, and some of them are localized within nucleus such as SIRT1, 6 and 7, where they target histones, act as transcription regulators and are involved in different processes including promotion of genome stability (SIRT6 and SIRT7), repair (SIRT7), gluconeogenesis and fatty acid oxidation (SIRT1) and regulation of triglyceride synthesis, NF-κB signalling and glucose homeostasis (SIRT6). On the contrary, SIRT3, 4 and 5 are active within mitochondria in which they improve the efficacy of electron transport chain (SIRT3 and SIRT4), activate the superoxide dismutase (SIRT5) and block glutamine entrance to tricarboxylic cycle (SIRT4) [167–169].

Knockout and overexpression studies of sirtuin genes widely report that these enzymes significantly extend lifespan, supporting the hypothesis that they are conserved mediators of longevity. This evidence has been developed in parallel with those demonstrating that the increase of acetylation levels with age of different histone residues, including histone 4 lysine 16 (H4K16) and H3K9 in worms, histone 4 lysine 16 (H4K16) in yeast, histone 3 lysine 56 (H3K56) in flies or H3K56 in mammals, is caused by a decreased activity of sirtuins [72, 162, 165, 167, 170–176]. It follows a heavy impact on chromatin transcription and, finally, in the regulation of different processes closely associated with ageing such as DNA repair and genome stability regulation, mitochondrial biogenesis, oxidative stress and telomere dysfunction.

Several environmental stimuli exert their effects on lifespan through changes of histone modifications. For example, dietary restriction (DR), such as a 30–40% reduction in the ad libitum levels of chow intake, widely demonstrated to extend lifespan and to delay age-related disease via sirtuin sensors by inducing the expression levels of SIRT1, SIRT3 and SIRT5 [177–179]. Reciprocally, a high-fat diet in mice and obesity in humans lead to the loss of SIRT1 [180–182]. Knocking out the mitochondrial SIRT3 prevents the protective effect of the dietary restriction [183]. Sirtuins exert their effects through several intracellular substrates with different physiological effects, including PPARγ coactivator 1α (PGC-1α), a key regulator of mitochondrial biogenesis and respiration, forkhead box protein O1 (FOXO1), a transcription factor sensor of the insulin signalling pathway, peroxisome proliferator-activated receptor alpha (PPARα), a major regulator of lipid metabolism in the liver, AMP-activated protein kinase (AMPK), playing a role in cellular energy homeostasis, glycolytic enzymes and one of their key transcriptional inducers, hypoxia-inducible factor 1-alpha (HIF-1α), considered the main transcriptional regulator of cellular and developmental responses to hypoxia [184–191].

Lastly, an interplay between histone modifications and DNA methylation exists in ageing. This interplay is mediated by some chromatin readers such as methyl CpG-binding protein 2 and methyl CpG, binding domain proteins, which recruit DNMTs, HDACs and transcriptional factors in a transcriptional silencing complex. An example is provided by the polycomb repressive complex 2 (PRC2), a family of polycomb (Pc) group proteins that plays a role in early embryonic development, stem cell differentiation and tumorigenesis.

In this context, SIRT1 is able to deacetylate DNMT1, whereby to high levels of SIRT1 expression, corresponding high levels of genomic DNA methylation. It follows that the decreased activity of SIRT1 can be responsible of the observed global loss of DNA methylation with age, thus impacting different intracellular processes including maintaining genome stability, apoptosis and DNA repair. Not by chance, recently, resveratrol, the best known flavonoid in red grapes and wine associated with cardiovascular diseases, has been shown to significantly increase long interspersed element-1 (LINE 1) methylation by modulating SIRT1 and DNMT1 activities in cellular models of oxidative stress [192].

7.4.3 miRNAs and Ageing

The first study, carried out to investigate the role of miRNAs in ageing, was conducted by De Lencastre et al. in *C. elegans*. Using deep sequencing and considering young, middle-aged, wild-type and long-lived daf-2 insulin signalling mutants, they identified 11 miRNAs that change with age and share homology with higher eukaryotes. The expression of some of these identified miRNAs declined with age, and in particular, let-7 showed the highest decrease with age. Furthermore, using knockout mutants, De Lencastre et al. contributed to define that these 11 miRNAs are not simply passive biomarkers but have an active regulatory role in ageing, as changes in their expression increase stress resistance and extend lifespan. Deepening the knowledge about their targets in the longevity, miRNA-mediated mechanism, insulin/IGF-1 signalling (IIS) and cell-cycle checkpoint pathways resulted as the mainly involved processes. The role of some of these detected miRNAs in both development and lifespan regulation prompted to investigate whether the influence of miRNAs on lifespan is a secondary aspect of their primary function on development. Evidence indicated that they regulate ageing independently of development [193].

The model organism *C. elegans* has also been studied to detect the role of miRNAs in the reduction of robustness, typical of ageing. Robustness is the ability to tolerate and resist to perturbations that affect the functionality of the system. In this regard, miR-34 showed an effect in increasing robustness by targeting the transcription factors DAF-16/FOXO. In fact, when upregulated in *C. elegans*, miR-34 stops the development and induces the adaptation to a lower metabolic state in order to avoid a stress-related damage, thus increasing robustness [194]. Therefore, by regulating transcriptional mechanisms, miRNAs may also control the reduction of robustness in human ageing.

Subsequently, the studies to identify age-related miRNAs have continued across different species.

In humans, for instance, the Baltimore Longitudinal Study of Ageing (BLSA) has been conducted in order to detect the serum miRNA profile in long-lived patients (76–92 years), compared with the short-lived subgroup (58–75 years). This study showed that 24 miRNAs were significantly upregulated and 73 were downregulated in the long-lived subgroup. The most upregulated was miR-373-5p, whereas the most downregulated was miR-15b-5p [195]. Moreover, a correlation between lifespan and the expression of some miRNAs (miR-211-5p, 374a-5p, 340-3p, 376c-3p, 5095, 1225-3p) has been observed. Interestingly, among the targets of these miR-NAs, there are ageing-related genes such as PARP1 (Poly ADP-Ribose Polymerase 1) involved in the processes of differentiation and proliferation and IGF1R (Insulin-Like Growth Factor 1 Receptor) and IGF2R (Insulin-Like Growth Factor 2 Receptor) involved in cell growth and survival control.

There are also some circulating miRNAs that are differentially expressed in human ageing such as miR-15b, miR-26a and miR-301a [102]. Some of these miR-NAs regulate the expression of genes that are involved in the apoptotic mechanisms. For instance, miRNAs with apoptotic targets are miR-26a that binds to BAK1 (BCL2 Antagonist/Killer 1), miR-24 and miR-181 that bind to BIM (Bcl-2 Interacting Mediator of Cell Death), miR-23 that targets BID (BH3 Interacting Domain Death Agonist), miR-23 that regulates CASP7 (Caspase-7) and NIX/ BNIP3L (NIP3-Like Protein X/BCL2 Interacting Protein 3 Like), miR-30 that binds to CASP3 (Caspase-3) and miR-27 that regulates APAF1 (Apoptotic Peptidase Activating Factor 1) [196].

Furthermore, several studies have been conducted in order to understand the role of miRNAs in inflammation, since it is well known that ageing is characterized by a systemic chronic inflammation that favours the manifestation of diseases and disabilities. Evidence showed that this inflammatory state is associated with the upregulation of some miRNAs, the so-called inflamma-miRs. The inflamma-miRs are involved in the responsiveness of neutrophils, macrophages, endothelial and immune cells by regulating inflammatory molecules such as Toll-Like Receptor-4 (let-7), NF-kB (miR-9), and Vascular Cell Adhesion Molecule 1 (miR-126). In particular, miRNAs involved in toll-like receptor and NF-kB pathways are upregulated in order to restrain the excessive inflammatory response [197].

In the past years, several researches have highlighted mitochondria impairment during ageing, characterized by autophagy decline, imbalance between mitochondrial fusion and fission and accumulation of protein aggregates that lead to dysfunctional mitochondria (reviewed in [198]). Therefore, miRNAs that target mitochondrial- or nuclear-encoded mitochondrial genes (mitomiRs) have been subjects of studies [197, 199]. A recent study showed that the mitomiRs miR-146a, miR-34a and miR-181a are overexpressed in replicative senescent human endothelial cells (HUVECs) compared with young cells, whereas their target Bcl-2 is downregulated [197]. Bcl-2 has anti-apoptotic and anti-oxidant functions, and therefore, its downregulation through the action of miRNAs causes increased oxidative stress and impaired functions. In addition, in silico analyses show an association between the mitomiR miR-146a, CHUK gene of NF-kB pathway and IRS1 of the insulin pathway [197]. In particular, some mitomiRs can reduce insulin secretion by targeting genes that encode for mitochondrial transporters [200]. Thus, the mitomiRs seem to be involved in NF-kB and insulin pathways, as observed for inflamma-miRs.

Parallel to researches focusing on ageing-related genes that are targeted by miR-NAs, several studies have been carried out in order to evaluate how these miRNAs are influenced by environmental factors. This aspect has been particularly investigated for circulating miRNAs. In this regard, researchers have profiled the expression of miRNAs in nine male monozygotic twins born between 1917 and 1927. As a result, 866 miRNAs have been detected in the plasma samples, and showed to have different levels in the two twins. Furthermore, comparing deceased twins with their living co-twin brothers, 34 miRNAs was more expressed in deceased twins, whereas 30 miRNAs had lower levels [201]. The differences of circulating miRNAs in identical twins, demonstrated a crucial role of environmental factors in miRNA-mediated life expectancy.

Proceeding in the analysis of environmental factors that influence ageing-related miRNAs, the role of diet has been emphasized. For instance, a study in mice has shown that miR-1, a significant miRNA in several age-related diseases, can be modulated by diet, since it was downregulated in the adipose tissue of mice fed with high-fat diet [202]. In humans, the expression of miR-22-3p in obese individuals leads to the downregulation of SIRT1, which modulates the endothelial cellular senescence. The upregulation of SIRT7 has been instead observed in slim individuals [203], and miR-19b could be a potential biomarker of polyunsaturated fatty acid intake, as its circulating levels increase after 8 weeks of fatty acid-enriched diet. Interestingly, miR-19b is downregulated in octogenarians, while centenarians and young people conserve the same levels of expression, suggesting that miRNAs influenced by diet could be used as biomarkers to predict longevity [202].

Regular exercise, as is well known, is one of the most important interventions to slow down the ageing process and maintain a longer healthy life. Recent studies have shown that exercise plays a role in the modification of miRNA profile. For instance, a significant buildup of miR-1, miR-133a, miR-133b and miR-181a caused by exercise has been detected in skeletal muscle biopsies taken from aged (70 ± 2 years) subjects, compared with young (29 ± 2 years) men. Among them, miR-1 downregulates the expression of the transcription factors Pax3 and Pax7, which are expressed during muscle development, in order to maintain the appropriate levels of satellite cells in the elderly skeletal muscle, hence contributing to a regular muscle growth and regeneration [204].

7.5 Epigenetic Modifications in Age-Related Diseases

Considering that epigenetic marks induce profound changes in the gene expression and contribute to the cellular and organismal phenotypic plasticity during lifetime, it is evident that dysregulations of epigenetic patterns may contribute to age-related diseases, including cancer, diabetes, cardiovascular and neurodegenerative diseases. Indeed, in all of these diseases, methylome-wide association studies (MWAS) have identified characteristic methylome signatures and brought to light as alteration in DNA methylation, and histone modifications are hallmarks often coincident with those observed in ageing, thus suggesting their importance as molecular marker prevention, prognosis and therapeutic approaches. Example of this overlapping come from the observation that DNA hypomethylation, prevalently at repetitive DNA elements, locus-specific hypermethylation of tumour suppressor genes (*p53*, *p21*, *p16*, *TIG1* and *RB1*), oncogenes (*cMYC* and *TERT*), genes involved in type 2 diabetes (*COX7A1*, *PRDX2*, *IRS1* and *KCNJ11*), genes involved in Alzheimer's disease (*APP*, *PS1* and *BACE1*) as well as histone hypo- and hypermethylation and/or a-acetylation generally occurring in the above diseases during ageing, thus causing aberrant gene expression and change to higher-order chromatin structure. It follows that epigenetic-based drugs which reverse aberrant DNA methylation or histone profiles through the modulation of DNMTs and histone-modifying enzymes may be considered effective in the assessment and development of epigenetic-based treatments [205].

The altered patterns of miRNA expression are also implicated in some age-related diseases as shown in Table 7.3. miRNAs are mostly involved in biological

Table 7.3 Effects of miRNAs and their targets in age-related phenotypes

miRNA	Target	Function	Disease	Reference
miR-145	*ERG, KLF5*	Inhibits proliferation and induces apoptosis	Downregulated in breast cancer Downregulated in atherosclerosis	Croce [319] Jovanović et al. [320]
miR-34	*CDK4, CDK6 CCNE2, EZF3, MET*	Induces apoptosis	Downregulated in breast cancer Upregulated in Alzheimer	Croce [319] Shah et al. [212]
miR-29	*TCL1. MCL1, DNMT*	Induces apoptosis	Downregulated in breast cancer	Croce [319]
Let-7	*RAS, MYC, HMGA2*	Induces apoptosis	Downregulated in breast and lung cancer	Croce [319]
miR-155	*MAF, SHIP1, MEF2A, NF-kB, TLR, AT1R*	Induces cell proliferation Reduces angiotensin II activity	Upregulated in breast and lung cancer Upregulated in aged muscle Expressed in atherosclerosis plaques	Croce [319] Jingjing et al. [210] Jovanović et al. [320]
miR-200c	*TRKB*	Induces cell proliferation	Upregulated in gastric cancer	Macha et al. [321]
miR-373	*IGF1R*	Involved in cell survival control	Downregulated in colon cancer	Macha et al. [321]
miRNA-199a-3p	*CD44*	Involved in cell adhesion and migration	Upregulated in gastric cancer	Macha et al. [321]
miR-21	*CDC25A, PTEN*	Induces cell proliferation	Upregulated in oesophageal cancer	Macha et al. [321]
miR-146a	*SOCS1, TLR4, NF-kB*	Induces cell growth and survival Reduces the expression of IL-1β and IL-6	Upregulated in gastric cancer Upregulated in aged muscle	Macha et al. [321] Jingjing et al. [210]

(continued)

Table 7.3 (continued)

miRNA	Target	Function	Disease	Reference
miR-148a	DNMT3B	Involved in development	Upregulated in gastric cancer	Macha et al. [321]
miR-1, miR-206, miR-133	MYOD1, MYOG, PAX7, HDAC4, FGF, PITX3, LPPR4, MYOCD	Induces myotube differentiation and reinnervation Induces dopaminergic neuron differentiation	Upregulated in injured muscle Downregulated in Parkinson Upregulated in coronary artery occlusion	Nakasa et al. [214] Williams et al. [322] Shah et al. [212] Schulte et al. [213]
miR-182	FOXO	Reduces myofibre atrophy	Upregulated in muscle atrophy	Hudson et al. [323]
miR-30a, miR-26a2	SYNJ2, NEGR1	Induces neurodegeneration	Upregulated in Parkinson	Shinde et al. [211]
miR-29a, miR-29b-1 and miR-9	BACE1	Induces the formation of senile plaques	Downregulated in Alzheimer	Shah et al. [212]
miR-106a and miR-106b	APP	Induces the formation of amyloid plaques	Downregulated in Alzheimer	Shah et al. [212]
miR-214	XBP1	Suppresses angiogenesis	Upregulated in heart dysfunction	Duan et al. [215]
miR-223	HMGCS1, SCARB1, SR-BI, SP3	Coordinates cholesterol homeostasis	Upregulated in hypercholesterolemia	Vickers et al. [324]
miR-19a	CCND1	Induces cell cycle arrest	Upregulated in LS disturbed flow	Kumar et al. [325]
miR-92a	KLF2, KLF4	Induces endothelial inflammation	Downregulated in LS and upregulated in OS disturbed flow	Kumar et al. [325]

processes such as cell division, apoptosis, intracellular signalling and cellular metabolism; therefore, the early stages of the research about the functions of miR-NAs in human diseases have been carried out to evaluate their role in cancer. The first disclosure was a down-expression of miR-15 and miR-16 among patients affected by B-cell chronic lymphocytic leukaemia [206]. Interestingly, previous studies have shown that miR-15 targets the integrin gene (ITGB5) that mediates cell adhesion and miR-16 binds to ATP5E (ATP Synthase Subunit Epsilon) and IGF-1 (Insulin-Like Growth Factor 1), suggesting a role in angiogenesis and metabolic dysfunction in cancer [207]. Generally, the altered expression of miRNAs in cancer occurs as a downregulation or upregulation. Evidence has also shown that some miRNAs are differentially expressed during various cancer progression stages, thus they are helpful to define diagnoses and for prognosis assessments. miRNAs can act as tumour suppressors or oncogenes through a translational and posttranscriptional regulation of tumorigenic elements. For instance, the oncogenic miR-125b has high levels of expression in pancreatic cancer, and it promotes cell proliferation, inhibiting p53-dependent apoptosis [208]. Thus, these cancer-related miRNAs can be employed to improve anticancer therapeutic strategies.

One of the most significant disabilities associated with ageing is sarcopenia. Individuals with sarcopenia show loss of muscle mass and function, frequent falls and balance problems. miRNAs that operate in muscle mechanisms such as myogenesis and muscle homeostasis are termed myomiRs [209]. Some of these miRNAs promote the regeneration of injured muscle, and they are upregulated during exercise, suggesting the importance of physical activity to prevent sarcopenia. miRNAs are also able to modulate the production of proinflammatory cytokines in aged muscle. For instance, miR-146a reduces the expression of IL-1β and IL-6 while miR-155 reinforces the translation of TNF-α in metabolic syndromes [210]. Based on these evidences, the regulation of inflammatory elements through miRNAs can drive the future research for diagnosis and treatment of sarcopenia.

In addition, neurodegenerative diseases are spread in elderly population, especially Parkinson and Alzheimer diseases. Recent studies have suggested that miR-30a, together with miR-26a2, may contribute to the susceptibility of Parkinson, in fact they are highly expressed in the blood cells of patients compared to control subjects [211]. Shah and colleagues have reported that miR-133, involved in the differentiation of dopaminergic neurons, is downregulated in Parkinson disease patients [212]. Interestingly, this miRNA is upregulated in coronary artery occlusion and in injured muscle [213, 214]. The future perspectives may be addressed to the better detection of miRNA sets in neurological tissues in order to understand their way of action more efficiently.

Moreover, several evidences indicate that altered miRNA functions contribute to cardiovascular diseases with a major incidence in aged individuals. For example, miR-214 is upregulated in heart dysfunction and hypertrophy, and it attenuates the cardiac angiogenesis, by decreasing its target XBP1, a transcription factor of unfolded protein response [215]. Some miRNAs have a role in the formation of the atherosclerotic plaque; therefore, they can be used for the prediction of atherosclerosis. Myocardial miRNAs are also significant. One specific heart miRNA is miR-1 that increases 12 hours after coronary artery occlusion; therefore, its level can be predictive of myocardium damage [216]. Another miRNA released during myocardial infarction is miR-208, and it seems to have an overlapping time of release with troponin [216, 217]. Based on these evidences, miRNAs could be used in addition to usual biomarkers in future diagnosis approaches.

In conclusion, to understand the mechanisms underlying epigenetic modifications, it has been (and will certainly be more and more) important to elucidate how environmental and genetic factors do cooperate to determine the ageing process and ageing-related phenotypes and diseases. On the other hand, data are accumulating which show that epigenetic modifications may represent important tools to monitor the rate and the quality of ageing, or to warn for the onset of age-related diseases.

References

1. Barros SP, Offenbacher S. Epigenetics: connecting environment and genotype to phenotype and disease. J Dent Res. 2009;88:400–8.
2. Illingworth RS, Bird AP. CpG islands—'a rough guide'. FEBS Lett. 2009;583:1713–20.
3. Kanherkar RR, Bhatia-Dey N, Csoka AB. Epigenetics across the human lifespan. Front Cell Dev Biol. 2014;2:49.
4. Schübeler D. Function and information content of DNA methylation. Nature. 2015;517: 321–6.
5. Luo G-Z, Blanco MA, Greer EL, He C, Shi Y. DNA N(6)-methyladenine: a new epigenetic mark in eukaryotes? Nat Rev Mol Cell Biol. 2015;16:705–10.
6. Wu TP, Wang T, Seetin MG, Lai Y, et al. DNA methylation on N(6)-adenine in mammalian embryonic stem cells. Nature. 2016;532:329–33.
7. Sánchez-Romero MA, Cota I, Casadesús J. DNA methylation in bacteria: from the methyl group to the methylome. Curr Opin Microbiol. 2015;25:9–16.
8. Lister R, Pelizzola M, Dowen RH, et al. Human DNA methylomes at base resolution show widespread epigenomic differences. This was the first report of a human methylome at single-base resolution. Nature. 2009;462:315–22.
9. Pinney SE. Mammalian non-CpG methylation: stem cells and beyond. Biology. 2014;3(4):739–51.
10. Patil V, Ward RL, Hesson LB. The evidence for functional non-CpG methylation in mammalian cells. Epigenetics. 2014;9(6):823–8.
11. Deaton AM, Bird A. CpG islands and the regulation of transcription. Genes Dev. 2011;25:1010–22.
12. Goldberg AD, Allis CD, Bernstein E. Epigenetics: a landscape takes shape. Cell. 2007;128:635–8.
13. Gardiner-Garden M, Frommer M. CpG islands in vertebrate genomes. J Mol Biol. 1987;196:261–82.
14. Moore LD, Le T, Fan G. DNA methylation and its basic function. Neuropsychopharmacology. 2013;38:23–38.
15. Norouzitallab P, Baruah K, Vanrompay D, Bossier P. Can epigenetics translate environmental cues into phenotypes? Sci Total Environ. 2018;647:1281–93.
16. Saxonov S, Berg P, Brutlag DL. A genome-wide analysis of CpG dinucleotides in the human genome distinguishes two distinct classes of promoters. Proc Natl Acad Sci U S A. 2006;103(5):1412–7.
17. Larsen F, Gundersen G, Lopez R, Prydz H. CpG islands as gene markers in the human genome. Genomics. 1992;13:1095–107.
18. Zhu J, He F, Hu S, Yu J. On the nature of human housekeeping genes. Trends Genet. 2008;24:481–4.
19. Maunakea AK, Nagarajan RP, Bilenky M, Ballinger TJ, et al. Conserved role of intragenic DNA methylation in regulating alternative promoters. Nature. 2010;466:253–7.
20. Jones PA. Functions of DNA methylation: islands, start sites, gene bodies and beyond. Nat Rev Genet. 2012;13:484–92.
21. Okano M, Bell DW, Haber DA, Li E. DNA methyltransferases Dnmt3a and Dnmt3b are essential for de novo methylation and mammalian development. Cell. 1999;99:247–57.
22. Lyko F. The DNA methyltransferase family: a versatile toolkit for epigenetic regulation. Nat Rev Genet. 2018;19:81–92.
23. Dan J, Chen T. Genetic studies on mammalian DNA methyltransferases. Adv Exp Med Biol. 2016;945:123–50.
24. Branco MR, Ficz G, Reik W. Uncovering the role of 5-hydroxymethylcytosine in the epigenome. Nat Rev Genet. 2011;13:7–13.
25. Guo F, Li X, Liang D, Li T, et al. Active and passive demethylation of male and female pronuclear DNA in the mammalian zygote. Cell Stem Cell. 2014;15:447–59.

26. Saitou M, Kagiwada S, Kurimoto K. Epigenetic reprogramming in mouse pre-implantation development and primordial germ cells. Development. 2012;139:15–31.
27. Sadakierska-Chudy A, Kostrzewa RM, Filip M. A comprehensive view of the epigenetic landscape part I: DNA methylation, passive and active DNA demethylation pathways and histone variants. Neurotox Res. 2015;27:84–97.
28. Penn NW, Suwalski R, O'Riley C, Bojanowski K, Yura R. The presence of 5-hydroxymethylcytosine in animal deoxyribonucleic acid. Biochem J. 1972;126:781–90.
29. Naveh-Many T, Cedar H. Active gene sequences are undermethylated. Proc Natl Acad Sci U S A. 1981;78(7):4246–50.
30. Waechter DE, Baserga R. Effect of methylation on expression of microinjected genes. Proc Natl Acad Sci U S A. 1982;79:1106–10.
31. Hotchkiss RD. The quantitative separation of purines, pyrimidines, and nucleosides by paper chromatography. J Biol Chem. 1948;175(1):315–32.
32. Griffith JS, Mahler HR. DNA ticketing theory of memory. Nature. 1969;223:580–2.
33. Tahiliani M, Koh KP, Shen Y, Pastor WA, et al. Conversion of 5-methylcytosine to 5-hydroxymethylcytosine in mammalian DNA by MLL partner TET1. Science. 2009;324:930–5.
34. Koh KP, Yabuuchi A, Rao S, et al. Tet1 and Tet2 regulate 5-hydroxymethylcytosine production and cell lineage specification in mouse embryonic stem cells. Cell Stem Cell. 2011;8:200–13.
35. Wu X, Zhang Y. TET-mediated active DNA demethylation: mechanism, function and beyond. Nat Rev Genet. 2017;18:517–34.
36. Jones PA. Effects of 5-azacytidine and its 2'-deoxyderivative on cell differentiation and DNA methylation. Pharmacol Ther. 1985;28:17–27.
37. Jones PA. Altering gene expression with 5-azacytidine. Cell. 1985;40:485–6.
38. Keshet I, Yisraeli J, Cedar H. Effect of regional DNA methylation on gene expression. Proc Natl Acad Sci U S A. 1985;82:2560–4.
39. Yisraeli J, Frank D, Razin A, Cedar H. Effect of in vitro DNA methylation on beta-globin gene expression. Proc Natl Acad Sci. 1988;85:4638–42.
40. Kass SU, Goddard JP, Adams RL. Specific methylation of vector sequences inhibits transcription from the SV40 early promoter. Biochem Soc Trans. 1993;21:9.
41. Seelan RS, Mukhopadhyay P, Pisano MM, Greene RM. Effects of 5-Aza-2'-deoxycytidine (decitabine) on gene expression. Drug Metab Rev. 2018;50:193–207.
42. Yan X, Ehnert S, Culmes M, et al. 5-azacytidine improves the osteogenic differentiation potential of aged human adipose-derived mesenchymal stem cells by DNA demethylation. PLoS One. 2014;9:90846.
43. Clouaire T, Stancheva I. Methyl-CpG binding proteins: specialized transcriptional repressors or structural components of chromatin? Cell Mol Life Sci. 2008;65:1509–22.
44. Sasai N, Defossez PA. Many paths to one goal? The proteins that recognize methylated DNA in eukaryotes. Int J Dev Biol. 2009;53:323–34.
45. Singal R, Ginder GD. DNA methylation. Blood. 1999;93:4059–70.
46. Watt F, Molloy PL. Cytosine methylation prevents binding to DNA of a HeLa cell transcription factor required for optimal expression of the adenovirus major late promoter. Genes Dev. 1988;2:1136–43.
47. Boyes J, Bird A. DNA methylation inhibits transcription indirectly via a methyl-CpG binding protein. Cell. 1991;64:1123–34.
48. Yin Y, Morgunova E, Jolma A, et al. Impact of cytosine methylation on DNA binding specificities of human transcription factors. Science. 2017;356(6337).
49. Huang B, Jiang C, Zhang R. Epigenetics: the language of the cell? Epigenomics. 2014;6:73–88.
50. Allis CD, Jenuwein T. The molecular hallmarks of epigenetic control. Nat Rev Genet. 2016;17:487–500.
51. Bellizzi D, D'aquila P, Scafone T, Giordano M, Riso V, Riccio A, Passarino G. The control region of mitochondrial DNA shows an unusual CpG and non-CpG methylation pattern. DNA Res. 2013;20:537–47.

52. Ghosh S, Sengupta S, Scaria V. Comparative analysis of human mitochondrial methylomes shows distinct patterns of epigenetic regulation in mitochondria. Mitochondrion. 2014;18:58–62.
53. Iacobazzi V, Castegna A, Infantino V, Andria G. Mitochondrial DNA methylation as a next-generation biomarker and diagnostic tool. Mol Genet Metab. 2013;110:25–34.
54. Shock LS, Thakkar PV, Peterson EJ, et al. DNA TSM methyltransferase 1, cytosine methylation, and cytosine hydroxymethylation in mammalian mitochondria. Proc Natl Acad Sci U S A. 2011;108(9):3630–5.
55. Rebelo AP, Williams S, Moraes CT. In vivo methylation of mtDNA reveals the dynamics of protein-mtDNA interactions. Nucleic Acids Res. 2009;37:6701–15.
56. Cutter AR, Hayes JJ. A brief review of nucleosome structure. FEBS Lett. 2015;589(20 Pt A):2914–22.
57. Margueron R, Reinberg D. Chromatin structure and the inheritance of epigenetic information. Nat Rev Genet. 2010;11:285–96.
58. Quina AS, Buschbeck M, Di Croce L. Chromatin structure and epigenetics. Biochem Pharmacol. 2006;72:1563–9.
59. Woodcock CL, Ghosh RP. Chromatin higher-order structure and dynamics. Cold Spring Harbor Perspect Biol. 2010;2:596.
60. Fazary AE, Ju YH, Abd-Rabboh HSM. How does chromatin package DNA within nucleus and regulate gene expression? Int J Biol Macromol. 2017;101:862–81.
61. Swygert SG, Peterson CL. Chromatin dynamics: interplay between remodeling enzymes and histone modifications. Biochim Biophys Acta. 2014;1839:728–36.
62. Bannister AJ, Kouzarides T. Regulation of chromatin by histone modifications. Cell Res. 2011;21:381–95.
63. Lawrence M, Daujat S, Schneider R. Lateral thinking: how histone modifications regulate gene expression. Trends Genet. 2016;32:42–56.
64. Zhang T, Cooper S, Brockdorff N. The interplay of histone modifications—writers that read. EMBO Rep. 2015;16:1467–81.
65. Wang R, Xin M, Li Y, Zhang P, Zhang M. The functions of histone modification enzymes in cancer. Curr Protein Pept Sci. 2016;17:438–45.
66. Kouzarides T. Chromatin modifications and their function. Cell. 2007;128(4):693–705.
67. Tessarz P, Kouzarides T. Histone core modifications regulating nucleosome structure and dynamics. Nat Rev Mol Cell Biol. 2014;15:703–8.
68. Dimitrova E, Turberfield AH, Klose RJ. Histone demethylases in chromatin biology and beyond. EMBO Rep. 2015;16:1620–39.
69. Gelato KA, Fischle W. Role of histone modifications in defining chromatin structure and function. Biol Chem. 2008;389:353–63.
70. Kiefer CM, Hou C, Little JA, Dean A. Epigenetics of beta-globin gene regulation. Mutat Res. 2008;647:68–76.
71. Bártová E, Krejcí J, Harnicarová A, Galiová G, Kozubek S. Histone modifications and nuclear architecture a review. J Histochem Cytochem. 2008;56(8):711–21.
72. Maleszewska M, Mawer JSP, Tessarz P. Histone modifications in ageing and lifespan regulation. Curr Mol Biol Rep. 2016;2(1):26–35.
73. Marmorstein R, Zhou MM. Writers and readers of histone acetylation: structure, mechanism, and inhibition. Cold Spring Harb Perspect Biol. 2014;6:18762.
74. Seto E, Yoshida M. Erasers of histone acetylation: the histone deacetylase enzymes. Cold Spring Harb Perspect Biol. 2014;6:18713.
75. Voss AK, Thomas T. Histone lysine and genomic targets of histone acetyltransferases in mammals. Bioessays. 2018;40(10):e1800078.
76. Wapenaar H, Dekker FJ. Histone acetyltransferases: challenges in targeting bi-substrate enzymes. Clin Epigenet. 2016;8:59.
77. Kirkland JG, Raab JR, Kamakaka RT. TFIIIC bound DNA elements in nuclear organization and insulation. Biochim Biophys Acta Gene Regul Mech. 2013;1829:418–24.

78. Marmorstein R, Roth SY. Histone acetyltransferases: function, structure, and catalysis. Curr Opin Genet Dev. 2001;11:155–61.
79. Greer EL, Shi Y. Histone methylation: a dynamic mark in health, disease and inheritance. Nat Rev Genet. 2012;13:343–57.
80. Smith BC, Denu JM. Chemical mechanisms of histone lysine and arginine modifications. Biochim Biophys Acta. 2009;1789:45–57.
81. Richon VM, Johnston D, Sneeringer CJ, et al. Chemogenetic analysis of human protein methyltransferases. Chem Biol Drug Des. 2011;78:199–210.
82. Jahan S, Davie JR. Protein arginine methyltransferases (PRMTs): role in chromatin organization. Adv Biol Regul. 2015;57:173–84.
83. Kimura H. Histone modifications for human epigenome analysis. J Hum Genet. 2013;58:439–45.
84. Hino S, Kohrogi K, Nakao M. Histone demethylase LSD1 controls the phenotypic plasticity of cancer cells. Cancer Sci. 2016;107:1187–92.
85. Shi Y, Lan F, Matson C, Mulligan P. Histone demethylation mediated by the nuclear amine oxidase homolog LSD1. J Chem Soc Faraday Trans. 1994;90:533–9.
86. Anand R, Marmorstein R. Structure and mechanism of lysine-specific demethylase enzymes. J Biol Chem. 2007;282:35425–9.
87. Nan X, Ng HH, Johnson CA, Laherty CD, et al. Transcriptional repression by the methyl-CpG-binding protein MeCP2 involves a histone deacetylase complex. Nature. 1998;393:386–9.
88. Kondo Y. Epigenetic cross-talk between DNA methylation and histone modifications in human cancers. Yonsei Med J. 2009;50:455–63.
89. Wang J, Hevi S, Kurash JK, Lei H, Gay F, et al. The lysine demethylase LSD1 (KDM1) is required for maintenance of global DNA methylation. Nat Genet. 2009;41:125–9.
90. Rose NR, Klose RJ. Understanding the relationship between DNA methylation and histone lysine methylation. Biochim Biophys Acta. 2014;1839:1362–72.
91. Du J, Johnson LM, Jacobsen SE, Patel DJ. DNA methylation pathways and their crosstalk with histone methylation. Nat Rev Mol Cell Biol. 2015;16:519–32.
92. Fiannaca A, La Rosa M, La Paglia L, Rizzo R, Urso A. NRC: non-coding RNA classifier based on structural features. BioData Min. 2017;10:27.
93. Lee RC, Feinbaum RL, Ambros V. The *C. elegans* heterochronic gene lin-4 encodes small RNAs with antisense complementarity to lin-14. Cell. 1993;75:843–54.
94. Lee H, Han S, Kwon CS, Lee D. Biogenesis and regulation of the let-7 miRNAs and their functional implications. Protein Cell. 2016;7:100–13.
95. Sheng P, Fields C, Aadland K, Wei T, Kolaczkowski O, Gu T, et al. Dicer cleaves 5'-extended microRNA precursors originating from RNA polymerase II transcription start sites. Nucleic Acids Res. 2018;46:5737–52.
96. Young-Kook K, Boseon K, Narry Kim V. Re-evaluation of the roles of DROSHA, Exportin 5, and DICER in microRNA biogenesis. Proc Natl Acad Sci U S A. 2016;113(13): E1881–9.
97. Zhang F, Wang D. The pattern of microRNA binding site distribution. Genes. 2017;8:296.
98. Moretti F, Thermann R, Hentze MW. Mechanism of translational regulation by miR-2 from sites in the 5' untranslated region or the open reading frame. RNA. 2010;16:2493–502.
99. Denzler R, McGeary SE, Title AC, Agarwal V, Bartel DP, Stoffel M. Impact of MicroRNA levels, target-site complementarity, and cooperativity or competing endogenous RNA-regulated gene expression. Mol Cell. 2016;64:565–79.
100. Webster MW, Stowell JAW, Tang TTL, Passmore LA. Analysis of mRNA deadenylation by multi-protein complexes. Methods. 2017;126:95–104.
101. Sohel MH. Extracellular/circulating microRNAs: release mechanisms, functions and challenges. Achievem Life Sci. 2016;10(2):175–86.
102. Jung HJ, Suh Y. Circulating miRNAs in ageing and ageing-related diseases. J Genet Genomics. 2014;41:465–72.
103. Jose AM. Movement of regulatory RNA between animal cells. Genesis. 2015;53:395–416.

104. Hayashi T, Hoffman MP. Exosomal microRNA communication between tissues during organogenesis. RNA Biol. 2017;14:1683–9.
105. Bayraktar R, Van Roosbroeck K, Calin GA. Cell-to-cell communication: microRNAs as hormones. Mol Oncol. 2017;11:1673–86.
106. Johnson FB, Sinclair DA, Guarente L. Molecular biology of aging. Cell. 1999;96:291–302.
107. Kirkwood TBL. Understanding the odd science of aging. Cell. 2005;120:437–47.
108. Sebastiani P, Solovieff N, Dewan AT, Walsh KM, et al. Genetic signatures of exceptional longevity in humans. PLoS One. 2012;7:29848.
109. Montesanto A, Dato S, Bellizzi D, Rose G, Passarino G. Epidemiological, genetic and epigenetic aspects of the research on healthy ageing and longevity. Immun Ageing. 2012;9:6.
110. D'Aquila P, Rose G, Bellizzi D, Passarino G. Epigenetics and aging. Maturitas. 2013;74:130–6.
111. López-Otín C, Blasco MA, Partridge L, Serrano M, Kroemer G. The hallmarks of aging. Cell. 2013;153:1194–217.
112. Pal S, Tyler JK. Epigenetics and aging. Sci Adv. 2016;2:1600584.
113. Li Y, Tollefsbol TO. Age-related epigenetic drift and phenotypic plasticity loss: implications in prevention of age-related human diseases. Epigenomics. 2016;8:1637–51.
114. Jones MJ, Goodman SJ, Kobor MS. DNA methylation and healthy human aging. Aging Cell. 2015;14:924–32.
115. Zampieri M, Ciccarone F, Calabrese R, et al. Reconfiguration of DNA methylation in aging. Mech Ageing Dev. 2015;151:60–70.
116. Guarasci F, D'Aquila P, Mandalà M, Garasto S, et al. Aging and nutrition induce tissue-specific changes on global DNA methylation status in rats. Mech Ageing Dev. 2018;174:47–54.
117. Amodio N, D'Aquila P, Passarino G, et al. Epigenetic modifications in multiple myeloma: recent advances on the role of DNA and histone methylation. Expert Opin Ther Targets. 2017;21:91–101.
118. Martin GM. Epigenetic drift in aging identical twins. Proc Natl Acad Sci U S A. 2005;102:10413–4.
119. Lipman T, Tiedje LB. Epigenetic differences arise during the lifetime of monozygotic twins. Am J Matern Nurs. 2006;31:204.
120. Kaminsky ZA, Tang T, Wang SC, et al. DNA methylation profiles in monozygotic and dizygotic twins. Nat Genet. 2009;41:240–5.
121. Bell JT, Spector TD. A twin approach to unraveling epigenetics. Trends Genet. 2011;27:116–25.
122. Tan Q, Christiansen L, Thomassen M, et al. Twins for epigenetic studies of human aging and development. Ageing Res Rev. 2013;12:182–7.
123. Mendelsohn AR, Larrick JW. Epigenetic drift is a determinant of mammalian lifespan. Rejuvenation Res. 2017;20:430–6.
124. Slieker RC, van Iterson M, Luijk R, et al. Age-related accrual of methylomic variability is linked to fundamental ageing mechanisms. Genome Biol. 2016;7:191.
125. Issa JP. Aging and epigenetic drift: a vicious cycle. J Clin Invest. 2014;124:24–9.
126. Teschendorff AE, Menon U, Gentry-Maharaj A, et al. Age-dependent DNA methylation of genes that are suppressed in stem cells is a hallmark of cancer. Genome Res. 2010;20:440–6.
127. Bacalini MG, D'Aquila P, Marasco E, Nardini C, Montesanto A, Franceschi C, Passarino G, Garagnani P, Bellizzi D. The methylation of nuclear and mitochondrial DNA in ageing phenotypes and longevity. Mech Ageing Dev. 2017;165:156–61.
128. Rakyan VK, Down TA, Maslau S, Andrew T, et al. Human aging-associated DNA hypermethylation occurs preferentially at bivalent chromatin domains. Genome Res. 2010;20:434–9.
129. Bell JT, Tsai PC, Yang TP, et al. Epigenome-wide scans identify differentially methylated regions for age and age-related phenotypes in a healthy ageing population. PLoS Genet. 2012;8(4):e1002629.
130. Ashapkin VV, Kutueva LI, Vanyushin BF. Aging as an epigenetic phenomenon. Curr Genomics. 2017;18:385–407.
131. Zhang Z, Deng C, Lu Q, Richardson B. Age-dependent DNA methylation changes in the ITGAL (CD11a) promoter. Mech Ageing Dev. 2002;123:1257–68.

132. Vijg J, Dollé ME. Genome instability: cancer or aging? Mech Ageing Dev. 2007;128:466–8.
133. Bollati V, Schwartz J, Wright R, et al. Decline in genomic DNA methylation through aging in a cohort of elderly subjects. Mech Ageing Dev. 2009;130:234–9.
134. Jintaridth P, Mutirangura A. Distinctive patterns of age-dependent hypomethylation in interspersed repetitive sequences. Physiol Genomics. 2010;41:194–200.
135. Wei L, Liu B, Tuo J, Shen D, et al. Hypomethylation of the IL17RC promoter associates with age-related macular degeneration. Cell Rep. 2012;2(5):1151–8.
136. Heyn H, Li N, Ferreira HJ, et al. Distinct DNA methylomes of newborns and centenarians. Proc Natl Acad Sci. 2012;109:10522–7.
137. Bellizzi D, D'aquila P, Giordano M, Montesanto A, Passarino G. Global DNA methylation levels are modulated by mitochondrial DNA variants. Epigenomics. 2012;4:17–27.
138. Bellizzi D, D'Aquila P, Montesanto A, Corsonello A, et al. Global DNA methylation in old subjects is correlated with frailty. Age. 2012;34:169–79.
139. D'Aquila P, Bellizzi D, Passarino G. rRNA-gene methylation and biological aging. Aging (Albany NY). 2018;10:7–8.
140. D'Aquila P, Montesanto A, Mandalà M, Garasto S, et al. Methylation of the ribosomal RNA gene promoter is associated with aging and age-related decline. Aging Cell. 2017;16:966–75.
141. Bocklandt S, Lin W, Sehl ME, et al. Epigenetic predictor of age. PLoS One. 2011;6:14821.
142. Hannum G, Guinney J, Zhao L, Zhang L, Hughes G, Sadda S. Genome-wide methylation profiles reveal quantitative views of human aging rates. Mol Cell. 2013;49:359–67.
143. Horvath S. DNA methylation age of human tissues and cell types. Genome Biol. 2013;14(10):R115.
144. Marioni RE, Shah S, McRae AF, et al. DNA methylation age of blood predicts all-cause mortality in later life. Genome Biol. 2015;16:25.
145. Christiansen L, Lenart A, Tan Q, et al. DNA methylation age is associated with mortality in a longitudinal Danish twin study. Aging Cell. 2016;15:149–54.
146. Perna L, Zhang Y, Mons U, et al. Epigenetic age acceleration predicts cancer, cardiovascular, and all-cause mortality in a German case cohort. Clin Epigenet. 2016;8:64.
147. Levine ME, Lu AT, Quach A, et al. An epigenetic biomarker of aging for lifespan and healthspan. Aging (Albany NY). 2018;10:573–91.
148. Obata Y, Furusawa Y, Hase K. Epigenetic modifications of the immune system in health and disease. Immunol Cell Biol. 2015;93:226–32.
149. Huidobro C, Fernandez AF, Fraga MF. Aging epigenetics: causes and consequences. Mol Aspects Med. 2013;34:765–81.
150. Lillycrop KA, Burdge GC. Maternal diet as a modifier of offspring epigenetics. J Dev Orig Health Dis. 2015;6:88–95.
151. Vickers MH. Early life nutrition, epigenetics and programming of later life disease. Nutrients. 2014;6:2165–78.
152. Park JH, Kim SH, Lee MS, Kim MS. Epigenetic modification by dietary factors: Implications in metabolic syndrome. Mol Aspects Med. 2017;54:58–70.
153. Lumey LH, Stein AD, Kahn HS, van der Pal-de Bruin KM, Blauw GJ, Zybert PA, Susser ES. Cohort profile: the Dutch Hunger Winter families study. Int J Epidemiol. 2007;36:1196–204.
154. Stein AD, Pierik FH, Verrips GHW, Susser ES, Lumey LH. Maternal exposure to the Dutch famine before conception and during pregnancy: quality of life and depressive symptoms in adult offspring. Epidemiology. 2009;20:909–15.
155. D'Aquila P, Montesanto A, Guarasci F, Passarino G, Bellizzi D. Mitochondrial genome and epigenome: two sides of the same coin. Front Biosci (Landmark Ed). 2017;22:888–908.
156. D'Aquila P, Giordano M, Montesanto A, et al. Age-and gender-related pattern of methylation in the MT-RNR1 gene. Epigenomics. 2015;7:707–16.
157. Truong TP, Sakata-Yanagimoto M, Yamada M, et al. Influence of age-dependent decrease of DNA hydroxymethylation in human T cells. J Clin Exp Hematopathol. 2015;55:1–6.
158. Szulwach KE, Li X, Li Y, Song CX, et al. 5-hmC–mediated epigenetic dynamics during postnatal neurodevelopment and aging Keith. Nat Neurosci. 2012;14:1607–16.

159. Chouliaras L, van den Hove DL, Kenis G, et al. Age-related increase in levels of 5-hydroxymethylcytosine in mouse hippocampus is prevented by caloric restriction. Curr Alzheimer Res. 2012;9:536–44.
160. Dzitoyeva S, Chen H, Manev H. Effect of aging on 5-hydroxymethylcytosine in brain mitochondria. Neurobiol Aging. 2012;33:2881–91.
161. Kochmanski J, Marchlewicz EH, Cavalcante RG, Sartor MA, Dolinoy DC. Age-related epigenome-wide DNA methylation and hydroxymethylation in longitudinal mouse blood. Epigenetics. 2018;13(7):779–92.
162. Feser J, Tyler J. Chromatin structure as a mediator of aging. FEBS Lett. 2011;585: 2041–8.
163. Wang Y, Yuan Q, Xie L. Histone modifications in aging: the underlying mechanisms and implications. Curr Stem Cell Res Ther. 2018;13:125–35.
164. Dimauro T, David G. Chromatin modifications: the driving force of senescence and aging? Aging (Albany NY). 2009;1:182–90.
165. Das C, Tyler JK. Histone exchange and histone modifications during transcription and aging. Biochim Biophys Acta. 2013;1819:332–42.
166. McCauley BS, Dang W. Histone methylation and aging: lessons learned from model systems. Biochim Biophys Acta. 2014;1839:1454–62.
167. Giblin W, Skinner ME, Lombard DB. Sirtuins: guardians of mammalian healthspan. Trends Genet. 2014;30:271–86.
168. Poulose N, Raju R. Sirtuin regulation in aging and injury. Biochim Biophys Acta. 2015;1852:2442–55.
169. Wątroba M, Dudek I, Skoda M, Stangret A, Rzodkiewicz P, Szukiewicz D. Sirtuins, epigenetics and longevity. Ageing Res Rev. 2017;40:11–9.
170. Dang W, Steffen KK, Perry R, Dorsey JA, Johnson FB, Shilatifard A, Kaeberlein M, Kennedy BK, Berger SL. Histone H4 lysine-16 acetylation regulates cellular lifespan. Nature. 2009;459(7248):802–7.
171. Kanfi Y, Peshti V, Gil R, Naiman S, Nahum L, Levin E, Kronfeld-Schor N, Cohen HY. SIRT6 protects against pathological damage caused by diet-induced obesity. Aging Cell. 2010;9:162–73.
172. Kanfi Y, Naiman S, Amir G, Peshti V, Zinman G, Nahum L, Bar-Joseph Z, Cohen HY. The sirtuin SIRT6 regulates lifespan in male mice. Nature. 2012;483:218–21.
173. Mostoslavsky R, Chua KF, Lombard DB, et al. Genomic instability and aging-like phenotype in the absence of mammalian SIRT6. Cell. 2006;124:315–29.
174. Peleg S, Sananbenesi F, Zovoilis A, et al. Altered histone acetylation is associated with age-dependent memory impairment in mice. Science. 2010;328:753–6.
175. Herranz D, Cañamero M, Mulero F, Martinez-Pastor B, Fernandez-Capetillo O, Serrano M. Sirt1 improves healthy ageing and protects from metabolic syndrome-associated cancer syndrome. Nat Commun. 2010;1:3.
176. Grabowska W, Sikora E, Bielak-Zmijewska A. Sirtuins, a promising target in slowing down the ageing process. Biogerontology. 2017;18:447–76.
177. Cohen HY, Miller C, Bitterman KJ, Wall NR, Hekking B, Kessler B, Howitz KT, Gorospe M, de Cabo R, Sinclair DA. Calorie restriction promotes mammalian cell survival by inducing the SIRT1 deacetylase. Science. 2004;305:390–2.
178. Lombard DB, Alt FW, Cheng HL, et al. Mammalian Sir2 homolog SIRT3 regulates global mitochondrial lysine acetylation. Mol Cell Biol. 2007;27:8807–14.
179. Nakagawa H, Nuovo GJ, Zervos EE, Martin EW Jr, Salovaara R, Aaltonen LA, de la Chapelle A. Age-related hypermethylation of the 5' region of MLH1 in normal colonic mucosa is associated with microsatellite-unstable colorectal cancer development. Cancer Res. 2001;61:6991–5.
180. Pedersen SB, Ølholm J, Paulsen SK, Bennetzen MF, Richelsen B. Low Sirt1 expression, which is upregulated by fasting, in human adipose tissue from obese women. Int J Obes (Lond). 2008;32:1250–5.

181. Costa Cdos S, Hammes TO, Rohden F, Margis R, Bortolotto JW, Padoin AV, Mottin CC, Guaragna RM. SIRT1 transcription is decreased in visceral adipose tissue of morbidly obese patients with severe hepatic steatosis. Obes Surg. 2010;20:633–9.
182. Chalkiadaki A, Guarente L. Sirtuins mediate mammalian metabolic responses to nutrient availability. Nat Rev Endocrinol. 2012;8:287–96.
183. Someya S, Tanokura M, Weindruch R, Prolla TA, Yamasoba T. Effects of caloric restriction on age-related hearing loss in rodents and rhesus monkeys. Curr Aging Sci. 2010;3:20–5.
184. Motta MC, Divecha N, Lemieux M, Kamel C, Chen D, Gu W, Bultsma Y, McBurney M, Guarente L. Mammalian SIRT1 represses forkhead transcription factors. Cell. 2004;116:551–63.
185. Rodgers JT, Lerin C, Haas W, Gygi SP, Spiegelman BM, Puigserver P. Nutrient control of glucose homeostasis through a complex of PGC-1alpha and SIRT1. Nature. 2005;434: 113–8.
186. Gerhart-Hines Z, Rodgers JT, Bare O, Lerin C, Kim SH, Mostoslavsky R, Alt FW, Wu Z, Puigserver P. Metabolic control of muscle mitochondrial function and fatty acid oxidation through SIRT1/PGC-1alpha. EMBO J. 2007;26:1913–23.
187. Feige JN, Lagouge M, Canto C, Strehle A, Houten SM, Milne JC, Lambert PD, Mataki C, Elliott PJ, Auwerx J. Specific SIRT1 activation mimics low energy levels and protects against diet-induced metabolic disorders by enhancing fat oxidation. Cell Metab. 2008;8:347–58.
188. Fulco M, Cen Y, Zhao P, Hoffman EP, McBurney MW, Sauve AA, Sartorelli V. Glucose restriction inhibits skeletal myoblast differentiation by activating SIRT1 through AMPK-mediated regulation of Nampt. Dev Cell. 2008;14:661–73.
189. Purushotham A, Schug TT, Xu Q, Surapureddi S, Guo X, L. X. Hepatocyte-specific deletion of SIRT1 alters fatty acid metabolism and results in hepatic steatosis and inflammation. Cell Metab. 2009;9:327–38.
190. Hallows WC, Yu W, Denu JM. Regulation of glycolytic enzyme phosphoglycerate mutase-1 by Sirt1 protein-mediated deacetylation. J Biol Chem. 2012;287:3850–8.
191. Guarente L. Calorie restriction and sirtuins revisited. Genes Dev. 2013;27:2072–85.
192. Maugeri A, Barchitta M, Mazzone MG, Giuliano F, Basile G Agodi A. Resveratrol Modulates SIRT1 and DNMT functions and restores LINE-1 methylation levels in ARPE-19 cells under oxidative stress and inflammation. Int J Mol Sci. 2018;19:2118.
193. De Lencastre A, Pincus Z, Zhou K, Kato M, Lee SS, Slack FJ. MicroRNAs both promote and antagonize longevity in C. elegans. Curr Biol. 2010;20:2159–68.
194. Isik M, Blackwell TK, Berezikov E. MicroRNA mir-34 provides robustness to environmental stress response via the DAF-16 network in C. elegans. Sci Rep. 2016;6:36766.
195. Smith-Vikos T, Liu Z, Parsons C, Gorospe M, Ferrucci L, Gill TM, Slack FJ. A serum miRNA profile of human longevity: findings from the Baltimore Longitudinal Study of Ageing (BLSA). Ageing (Albany NY). 2016;8:2971–83.
196. Kulshreshtha R, Ferracin M, Wojcik SE, Garzon R, Alder H, Agosto-Perez FJ, Davuluri R, Liu CG, Croce CM, Negrini M, Calin GA, Ivan M. A microRNA signature of hypoxia. Mol Cell Biol. 2007;27:1859–67.
197. Rippo MR, Olivieri F, Monsurrò V, Prattichizzo F, Albertini MC, Procopio AD. MitomiRs in human inflamm-ageing: a hypothesis involving miR-181a, miR-34a and miR-146a. Exp Gerontol. 2014;56:154–63.
198. Rose G, Santoro A, Salvioli S. Mitochondria and mitochondria-induced signalling molecules as longevity determinants. Mech Ageing Dev. 2017;165:115–28.
199. Olivieri F, Rippo MR, Monsurrò V, Salvioli S, Capri M, Procopio AD, Franceschi C. MicroRNAs linking inflamm-ageing, cellular senescence and cancer. Ageing Res Rev. 2013;12:1056–68.
200. Baradan R, Hollander JM, Das S. Mitochondrial miRNAs in diabetes: just the tip of the iceberg. Can J Physiol Pharmacol. 2017;95:1156–62.
201. Wu S, Kim T-K, Wu X, Scherler K, Baxter D, Wang K, Krasnow RE, Reed T, Dai J. Circulating microRNAs and life expectancy among identical twins. Ann Human Genet. 2016;80:247–56.

202. Micó V, Berninches L, Tapia J, Daimiel L. NutrimiRAging: Micromanaging Nutrient Sensing Pathways through Nutrition to Promote Healthy Aging. Int J Mol Sci. 2017;18:915.
203. Kurylowicz A, Owczarz M, Polosak J, Jonas MI, Lisik W, Jonas M, Chmura A, Puzianowska-Kuznicka M. SIRT1 and SIRT7 expression in adipose tissues of obese and normal-weight individuals is regulated by microRNAs but not by methylation status. Int J Obes (Lond). 2016;40:1635–42.
204. Ultimo S, Zauli G, Martelli AM, Vitale M, McCubrey JA, Capitani S, Neri LM. Influence of physical exercise on microRNAs in skeletal muscle regeneration, aging and diseases. Oncotarget. 2018;9:17220–37.
205. Brunet A, Berger SL. Epigenetics of aging and aging-related disease. J Gerontol A Biol Sci Med Sci. 2014;69:S17–20.
206. Shenouda SK, Alahari SK. MicroRNA function in cancer: oncogene or a tumor suppressor? Cancer Metastasis Rev. 2009;28:369.
207. Calin GA, Dumitru CD, Shimizu M, et al. Frequent deletions and down-regulation of micro-RNA genes miR15 and miR16 at 13q14 in chronic lymphocytic leukemia. Proc Natl Acad Sci U S A. 2002;99:15524–9.
208. Jiang W, Li J, Zhang Z, Wang H, Wang Z. Epigenetic upregulation of alpha-synuclein in the rats exposed to methamphetamine. Eur J Pharmacol. 2014;745:243–8.
209. Brown DM, Goljanek-Whysall K. microRNAs: modulators of the underlying pathophysiology of sarcopenia? Ageing Res Rev. 2015;24:263–73.
210. Jingjing F, Xianjuan K, Yi Y, Ning C. MicroRNA-regulated proinflammatory cytokines in sarcopenia. Mediators Inflamm. 2016;2016:1438686.
211. Shinde S, Mukhopadhyay S, Mohsen G, Khoo SK. Biofluid-based microRNA biomarkers for Parkinson's disease: an overview and update. AIMS Med Sci. 2015;2:15–25.
212. Shah P, Cho SK, Thulstrup PW, Bjerrum MJ, Lee PH, Kang JH, Bhang YJ, Yang SW. MicroRNA biomarkers in neurodegenerative diseases and emerging nanosensors technology. J Mov Disorders. 2017;10:18–28.
213. Schulte C, Zeller T. microRNA-based diagnostics and therapy in cardiovascular disease—summing up the facts. Cardiovasc Diagn Ther. 2015;5:17–36.
214. Nakasa T, Ishikawa M, Shi M, Shibuya H, Adachi N, Ochi M. Acceleration of muscle regeneration by local injection of muscle-specific microRNAs in rat skeletal muscle injury model. J Cell Mol Med. 2010;14:2495–505.
215. Duan Q, Yang L, Gong W, Chaugai S, Wang F, Chen C, Wang P, Zou MH, Wang DW. MicroRNA-214 is upregulated in heart failure patients and suppresses XBP1-mediated endothelial cells angiogenesis. J Cell Physiol. 2015;230:1964–73.
216. Alavi-Moghaddam M, Chehrazi M, Alipoor SD, Mohammadi M, Baratloo A, Mahjoub MP, Movasaghi M, Garssen J, Adcock IM, Mortaz E. A preliminary study of microRNA-208b after acute myocardial infarction: impact on 6-month survival. Dis Markers. 2018;2018:2410451.
217. Schulte C, Ji X, Takahashi R, Hiura Y, et al. Plasma miR-208 as a biomarker of myocardial injury. Clin Chem. 2009;55(11):1944–9.
218. Reynolds LM, Taylor JR, Ding J, et al. Age-related variations in the methylome associated with gene expression in human monocytes and T cells. Nat Commun. 2014;5:5366.
219. McClay JL, Aberg KA, Clark SL, et al. A methylome-wide study of aging using massively parallel sequencing of the methyl-CpG-enriched genomic fraction from blood in over 700 subjects. Hum Mol Genet. 2014;23(5):1175–85.
220. Choi EK, Uyeno S, Nishida N, et al. Alterations of c-fos gene methylation in the processes of aging and tumorigenesis in human liver. Mutat Res. 1996;354(1):123–8.
221. Issa JP, Vertino PM, Boehm CD, et al. Switch from monoallelic to biallelic human IGF2 promoter methylation during aging and carcinogenesis. Proc Natl Acad Sci U S A. 1996;93(21):11757–62.
222. Ahuja N, Li Q, Mohan AL, Baylin SB, Issa JP. Aging and DNA methylation in colorectal mucosa and cancer. Cancer Res. 1998;58(23):5489–94.

223. Christensen BC, Houseman EA, Marsit CJ, et al. Aging and environmental exposures alter tissue-specific DNA methylation dependent upon CpG island context. PLoS Genet. 2009;5(8):e1000602.
224. Vidal AC, Benjamin Neelon SE, Liu Y. Maternal stress, preterm birth, and DNA methylation at imprint regulatory sequences in humans. Genet Epigenet. 2014;6:37–44.
225. Nakagawa H, Nuovo GJ, Zervos EE, et al. Age-related hypermethylation of the 5' region of MLH1 in normal colonic mucosa is associated with microsatellite-unstable colorectal cancer development. Cancer Res. 2001;61(19):6991–5.
226. Matsubayashi H, Sato N, Brune K, et al. Age- and disease-related methylation of multiple genes in nonneoplastic duodenum and in duodenal juice. Clin Cancer Res. 2005;11(2 Pt 1):573–83.
227. Silva PN, Gigek CO, Leal MF, et al. Promoter methylation analysis of SIRT3, SMARCA5, HTERT and CDH1 genes in aging and Alzheimer's disease. J Alzheimers Dis. 2008;13(2):173–6.
228. Madrigano J, Baccarelli A, Mittleman MA, et al. Aging and epigenetics: longitudinal changes in gene-specific DNA methylation. Epigenetics. 2012;7(1):63–70.
229. Rönn T, Poulsen P, Hansson O, et al. Age influences DNA methylation and gene expression of COX7A1 in human skeletal muscle. Diabetologia. 2008;51(7):1159–68.
230. Gaudet MM, Campan M, Figueroa JD, et al. DNA hypermethylation of ESR1 and PGR in breast cancer: pathologic and epidemiologic associations. Cancer Epidemiol Biomarkers Prev. 2009;18(11):3036–43.
231. Fujii H, Biel MA, Zhou W, et al. Methylation of the HIC-1 candidate tumor suppressor gene in human breast cancer. Oncogene. 1998;16(16):2159–64.
232. Cody DT, Huang Y, Darby CJ, et al. Differential DNA methylation of the p16 INK4A/CDKN2A promoter in human oral cancer cells and normal human oral keratinocytes. Oral Oncol. 1999;35(5):516–22.
233. Dammann R, Li C, Yoon JH, et al. Epigenetic inactivation of a RAS association domain family protein from the lung tumour suppressor locus 3p21.3. Nat Genet. 2000;25(3):315–9.
234. Virmani AK, Rathi A, Sathyanarayana UG, et al. Aberrant methylation of the adenomatous polyposis coli (APC) gene promoter 1A in breast and lung carcinomas. Clin Cancer Res. 2001;7(7):1998–2004.
235. Waki T, Tamura G, Sato M, Motoyama T. Age-related methylation of tumor suppressor and tumor-related genes: an analysis of autopsy samples. Oncogene. 2003;22(26):4128–33.
236. Sutherland KD, Lindeman GJ, Choong DY, et al. Differential hypermethylation of SOCS genes in ovarian and breast carcinomas. Oncogene. 2004;23(46):7726–33.
237. So K, Tamura G, Honda T, Homma N, et al. Multiple tumor suppressor genes are increasingly methylated with age in non-neoplastic gastric epithelia. Cancer Sci. 2006;97(11):1155–8.
238. Nishida N, Nagasaka T, Nishimura T, et al. Aberrant methylation of multiple tumor suppressor genes in aging liver, chronic hepatitis, and hepatocellular carcinoma. Hepatology. 2008;47(3):908–18.
239. Yuan Y, Qian ZR, Sano T, et al. Reduction of GSTP1 expression by DNA methylation correlates with clinicopathological features in pituitary adenomas. Mod Pathol. 2008;21(7):856–65.
240. Pilsner JR, Hall MN, Liu X, et al. Influence of prenatal arsenic exposure and newborn sex on global methylation of cord blood DNA. PLoS One. 2012;7(5):e37147.
241. Majumdar S, Chanda S, Ganguli B, et al. Arsenic exposure induces genomic hypermethylation. Environ Toxicol. 2010;25(3):315–8.
242. Guo X, Chen X, Wang J, et al. Multi-generational impacts of arsenic exposure on genome-wide DNA methylation and the implications for arsenic-induced skin lesions. Environ Int. 2018;119:250–63.
243. Kaushal A, Zhang H, Karmaus WJJ. Genome-wide DNA methylation at birth in relation to in utero arsenic exposure and the associated health in later life. Environ Health. 2017;16(1):50.
244. Cowley M, Skaar DA, Jima DD, et al. Effects of cadmium exposure on DNA methylation at imprinting control regions and genome-wide in mothers and newborn children. Environ Health Perspect. 2018;126(3):037003.

245. Hirao-Suzuki M, Takeda S, Kobayashi T, et al. Cadmium down-regulates apolipoprotein E (ApoE) expression during malignant transformation of rat liver cells: direct evidence for DNA hypermethylation in the promoter region of ApoE. J Toxicol Sci. 2018;43(9):537–43.

246. Virani S, Rentschler KM, Nishijo M, et al. DNA methylation is differentially associated with environmental cadmium exposure based on sex and smoking status. Chemosphere. 2016;145:284–90.

247. Wang TC, Song YS, Wang H, et al. Oxidative DNA damage and global DNA hypomethylation are related to folate deficiency in chromate manufacturing workers. J Hazard Mater. 2012;213–214:440–6.

248. Yang L, Xia B, Yang X, et al. Mitochondrial DNA hypomethylation in chrome plating workers. Toxicol Lett. 2016;243:1–6.

249. Lou J, Wang Y, Yao C, et al. Role of DNA methylation in cell cycle arrest induced by Cr (VI) in two cell lines. PLoS One. 2013;8(8):e71031.

250. Cardenas A, Rifas-Shiman SL, Godderis L, et al. Prenatal exposure to mercury: associations with global DNA methylation and hydroxymethylation in cord blood and in childhood. Environ Health Perspect. 2017;125(8):087022.

251. Cardenas A, Rifas-Shiman SL, Agha G, et al. Persistent DNA methylation changes associated with prenatal mercury exposure and cognitive performance during childhood. Sci Rep. 2017;7(1):288.

252. Goodrich JM, Basu N, Franzblau A, Dolinoy DC. Mercury biomarkers and DNA methylation among Michigan dental professionals. Environ Mol Mutagen. 2013;54(3):195–203.

253. Zhang X, Chen X, Weirauch MT, et al. Diesel exhaust and house dust mite allergen lead to common changes in the airway methylome and hydroxymethylome. Environ Epigenet. 2018;4(3):dvy020.

254. Ghosh K, Chatterjee B, Kanade SR. Lead induces the up-regulation of the protein arginine methyltransferase 5 possibly by its promoter demethylation. Biochem J. 2018;475(16):2653–66.

255. Liu X, Wu J, Shi W, et al. Lead induces genotoxicity via oxidative stress and promoter methylation of DNA repair genes in human lymphoblastoid TK6 cells. Med Sci Monitor. 2018;24:4295–304.

256. Cheong A, Johnson SA, Howald EC, et al. Gene expression and DNA methylation changes in the hypothalamus and hippocampus of adult rats developmentally exposed to bisphenol A or ethinyl estradiol: a CLARITY-BPA consortium study. Epigenetics. 2018;13(7):704–20.

257. Mostafavi N, Vermeulen R, Ghantous A, et al. Acute changes in DNA methylation in relation to 24 h personal air pollution exposure measurements: a panel study in four European countries. Environ Int. 2018;120:11–21.

258. Maghbooli Z, Hossein-Nezhad A, Adabi E, et al. Air pollution during pregnancy and placental adaptation in the levels of global DNA methylation. PLoS One. 2018;13(7):e0199772.

259. Nawrot TS, Saenen ND, Schenk J, et al. Placental circadian pathway methylation and in utero exposure to fine particle air pollution. Environ Int. 2018;114:231–41.

260. Li J, Zhu X, Yu K, et al. Exposure to polycyclic aromatic hydrocarbons and accelerated DNA methylation aging. Environ Health Perspect. 2018;126(6):067005.

261. Lee J, Kalia V, Perera F, et al. Prenatal airborne polycyclic aromatic hydrocarbon exposure, LINE1 methylation and child development in a Chinese cohort. Environ Int. 2017;99:315–20.

262. White AJ, Chen J, Teitelbaum SL, et al. Sources of polycyclic aromatic hydrocarbons are associated with gene-specific promoter methylation in women with breast cancer. Environ Res. 2016;145:93–100.

263. White N, Benton M, Kennedy D, et al. Accounting for cell lineage and sex effects in the identification of cell-specific DNA methylation using a Bayesian model selection algorithm. PLoS One. 2017;12(9):e0182455.

264. Prince C, Hammerton G, Taylor AE, et al. Investigating the impact of cigarette smoking behaviours on DNA methylation patterns in adolescence. Hum Mol Genet. 2019;28(1):155–65.

265. Witt SH, Frank J, Gilles M, et al. Impact on birth weight of maternal smoking throughout pregnancy mediated by DNA methylation. BMC Genomics. 2018;19(1):290.

266. Cole E, Brown TA, Pinkerton KE, et al. Perinatal exposure to environmental tobacco smoke is associated with changes in DNA methylation that precede the adult onset of lung disease in a mouse model. Inhal Toxicol. 2017;29(10):435–42.

267. Sziráki A, Tyshkovskiy A, Gladyshev VN. Global remodeling of the mouse DNA methylome during aging and in response to calorie restriction. Aging Cell. 2018;17(3):e12738.

268. Kim CH, Lee EK, Choi YJ, et al. Short-term calorie restriction ameliorates genomewide, age-related alterations in DNA methylation. Aging Cell. 2016;15(6):1074–81.

269. Chen PY, Ganguly A, Rubbi L, et al. Intrauterine calorie restriction affects placental DNA methylation and gene expression. Physiol Genomics. 2013;45(14):565–76.

270. Pauwels S, Ghosh M, Duca RC, et al. Maternal intake of methyl-group donors affects DNA methylation of metabolic genes in infants. Clin Epigenet. 2017;9:16.

271. Pauwels S, Duca C, Devlieger R, et al. Maternal methyl-group donor intake and global DNA (hydroxy)methylation before and during pregnancy. Nutrients. 2016;8(8):474.

272. Kok DE, Dhonukshe-Rutten R, Lute C, et al. The effects of long-term daily folic acid and vitamin B12 supplementation on genome-wide DNA methylation in elderly subjects. Clin Epigenet. 2015;7:121.

273. Fang MZ, Wang Y, Ai N, et al. Tea polyphenol (-)-epigallocatechin-3-gallate inhibits DNA methyltransferase and reactivates methylation-silenced genes in cancer cell lines. Cancer Res. 2003;63(22):7563–70.

274. Anderson CM, Gillespie SL, Thiele DK, et al. Effects of maternal vitamin D supplementation on the maternal and infant epigenome. Breastfeed Med. 2018;13(5):371–80.

275. Zappe K, Pointner A, Switzeny OJ, et al. Counteraction of oxidative stress by vitamin E affects epigenetic regulation by increasing global methylation and gene expression of MLH1 and DNMT1 dose dependently in Caco-2 cells. Oxid Med Cell Longev. 2018;2018:3734250.

276. Ramaiyan B, Talahalli RR. Dietary unsaturated fatty acids modulate maternal dyslipidemia-induced DNA methylation and histone acetylation in placenta and fetal liver in rats. Lipids. 2018;53(6):581–8.

277. Moody L, Chen H, Pan YX. Postnatal diet remodels hepatic DNA methylation in metabolic pathways established by a maternal high-fat diet. Epigenomics. 2017;9(11):1387–402.

278. Zhang Y, Wang H, Zhou D, et al. High-fat diet caused widespread epigenomic differences on hepatic methylome in rat. Physiol Genomics. 2015;47(10):514–23.

279. Nakatome M, Orii M, Hamajima M, et al. Methylation analysis of circadian clock gene promoters in forensic autopsy specimens. Legal Med (Tokyo). 2011;13(4):205–9.

280. Jiang W, Li J, Zhang Z, et al. Epigenetic upregulation of alpha-synuclein in the rats exposed to methamphetamine. Eur J Pharmacol. 2014;745:243–8.

281. Itzhak Y, Ergui I, Young JI. Long-term parental methamphetamine exposure of mice influences behavior and hippocampal DNA methylation of the offspring. Mol Psychiatry. 2015;20(2):232–9.

282. Jayanthi S, McCoy MT, Chen B, et al. Methamphetamine downregulates striatal glutamate receptors via diverse epigenetic mechanisms. Biol Psychiatry. 2014;76(1):47–56.

283. Anier K, Malinovskaja K, Aonurm-Helm A, et al. DNA methylation regulates cocaine-induced behavioral sensitization in mice. Neuropsychopharmacology. 2010;35(12):2450–61.

284. Pol Bodetto S, Carouge D, Fonteneau M, et al. Cocaine represses protein phosphatase-1Cβ through DNA methylation and methyl-CpG binding protein-2 recruitment in adult rat brain. Neuropharmacology. 2013;73:31–40.

285. Tian W, Zhao M, Li M, et al. Reversal of cocaine-conditioned place preference through methyl supplementation in mice: altering global DNA methylation in the prefrontal cortex. PLoS One. 2012;7:e33435.

286. Carouge D, Host L, Aunis D, et al. CDKL5 is a brain MeCP2 target gene regulated by DNA methylation. Neurobiol Dis. 2010;38(3):414–24.

287. Ajonijebu DC, Abboussi O, Mabandla MV, et al. Differential epigenetic changes in the hippocampus and prefrontal cortex of female mice that had free access to cocaine. Metab Brain Dis. 2018;33(2):411–20.

288. Ebrahimi G, Asadikaram G, Akbari H, et al. Elevated levels of DNA methylation at the OPRM1 promoter region in men with opioid use disorder. Am J Drug Alcohol Abuse. 2018;44(2):193–9.
289. Chorbov VM, Todorov AA, Lynskey MT, et al. Elevated levels of DNA methylation at the OPRM1 promoter in blood and sperm from male opioid addicts. J Opioid Manag. 2011;7(4):258–64.
290. McLaughlin P, Mactier H, Gillis C, et al. Increased DNA methylation of ABCB1, CYP2D6, and OPRM1 genes in newborn infants of methadone-maintained opioid-dependent mothers. J Pediatrics. 2017;190:180–184.e1.
291. Groh A, Rhein M, Buchholz V, et al. Epigenetic effects of intravenous diacetylmorphine on the methylation of POMC and NR3C1. Neuropsychobiology. 2017;75(4):193–9.
292. Groh A, Jahn K, Burkert A, et al. Epigenetic regulation of the promotor region of vascular endothelial growth factor-A and nerve growth factor in opioid-maintained patients. Eur Addict Res. 2017;23(5):249–59.
293. Watson CT, Szutorisz H, Garg P, et al. Genome-wide DNA methylation profiling reveals epigenetic changes in the rat nucleus accumbens associated with cross-generational effects of adolescent THC exposure. Neuropsychopharmacology. 2015;40(13):2993–3005.
294. Gerra MC, Jayanthi S, Manfredini M, et al. Gene variants and educational attainment in cannabis use: mediating role of DNA methylation. Transl Psychiatry. 2018;8(1):23.
295. Taqi MM, Bazov I, Watanabe H, et al. Prodynorphin CpG-SNPs associated with alcohol dependence: elevated methylation in the brain of human alcoholics. Addict Biol. 2011;16(3):499–509.
296. Philibert RA, Gunter TD, Beach SR, et al. MAOA methylation is associated with nicotine and alcohol dependence in women. Am J Med Genet Part B Neuropsychiatric Genet. 2008;147B(5):565–70.
297. Glahn A, Riera Knorrenschild R, Rhein M, et al. Alcohol-induced changes in methylation status of individual CpG sites, and serum levels of vasopressin and atrial natriuretic peptide in alcohol-dependent patients during detoxification treatment. Eur Addict Res. 2014;20(3):143–50.
298. Foroud T, Wetherill LF, Liang T, et al. Association of alcohol craving with alpha-synuclein (SNCA). Alcohol Clin Exp Res. 2007;31(4):537–45.
299. Ji C, Nagaoka K, Zou J, et al. Chronic ethanol-mediated hepatocyte apoptosis links to decreased TET1 and 5-hydroxymethylcytosine formation. FASEB J. 2019;33(2):1824–35.
300. Frey S, Eichler A, Stonawski V, et al. Prenatal alcohol exposure is associated with adverse cognitive effects and distinct whole-genome DNA methylation patterns in primary school children. Front Behav Neurosci. 2018;12:125.
301. Brückmann C, Islam SA, MacIsaac JL, et al. DNA methylation signatures of chronic alcohol dependence in purified CD3+ T-cells of patients undergoing alcohol treatment. Sci Rep. 2017;7(1):6605.
302. Weng JT, Wu LS, Lee CS, et al. Integrative epigenetic profiling analysis identifies DNA methylation changes associated with chronic alcohol consumption. Comput Biol Med. 2015;64:299–306.
303. Heberlein A, Muschler M, Frieling H, et al. Epigenetic down regulation of nerve growth factor during alcohol withdrawal. Addict Biol. 2013;18(3):508–10.
304. Brückmann C, Di Santo A, Karle KN, et al. Validation of differential GDAP1 DNA methylation in alcohol dependence and its potential function as a biomarker for disease severity and therapy outcome. Epigenetics. 2016;11(6):456–63.
305. Jasiewicz A, Rubiś B, Samochowiec J, et al. DAT1 methylation changes in alcohol-dependent individuals vs. controls. J Psychiatric Res. 2015;64:130–3.
306. Fiano V, Trevisan M, Fasanelli F, et al. Methylation in host and viral genes as marker of aggressiveness in cervical lesions: analysis in 543 unscreened women. Gynecol Oncol. 2018;151(2):319–26. pii: S0090-8258(18)31161-2
307. Jin J, Xu H, Wu R, et al. Aberrant DNA methylation profile of hepatitis B virus infection. J Med Virol. 2019;91(1):81–92.

308. Nunes JM, Furtado MN, de Morais Nunes ER, et al. Modulation of epigenetic factors during the early stages of HIV-1 infection in CD4+ T cells in vitro. Virology. 2018;523:41–51.
309. Gao X, Zhang Y, Brenner H. Associations of Helicobacter pylori infection and chronic atrophic gastritis with accelerated epigenetic ageing in older adults. Er J Cancer. 2017;117(8):1211–4.
310. Gupta H, Chaudhari S, Rai A, et al. Genetic and epigenetic changes in host ABCB1 influences malaria susceptibility to Plasmodium falciparum. PLoS One. 2017;12(4):e0175702.
311. Mehta D, Bruenig D, Carrillo-Roa T, et al. Genomewide DNA methylation analysis in combat veterans reveals a novel locus for PTSD. Acta Psychiatr Scand. 2017;136(5):493–505.
312. Peng H, Zhu Y, Strachan E, et al. Childhood trauma, DNA methylation of stress-related genes, and depression: findings from two monozygotic twin studies. Psychosom Med. 2018;80(7):599–608.
313. Wolf EJ, Logue MW, Morrison FG, et al. Posttraumatic psychopathology and the pace of the epigenetic clock: a longitudinal investigation. Psychol Med. 2018;13:1–10.
314. Song D, Qi W, Lv M, et al. Combined bioinformatics analysis reveals gene expression and DNA methylation patterns in osteoarthritis. Mol Med Rep. 2018;17(6):8069–78.
315. Hughes A, Smart M, Gorrie-Stone T, Het a. Socioeconomic position and DNA methylation age acceleration across the lifecourse. Am J Epidemiol. 2018;187(11):2346–54.
316. Swartz JR, Hariri AR, Williamson DE. An epigenetic mechanism links socioeconomic status to changes in depression-related brain function in high-risk adolescents. Mol Psychiatry. 2017;22(2):209–14.
317. Chan MA, Ciaccio CE, Gigliotti NM, et al. DNA methylation levels associated with race and childhood asthma severity. J Asthma. 2017;54(8):825–32.
318. Tehranifar P, Wu HC, Fan X, et al. Early life socioeconomic factors and genomic DNA methylation in mid-life. Epigenetics. 2013;8(1):23–7.
319. Croce CM. Causes and consequences of microRNA dysregulation in cancer. Nat Rev Genet. 2009;10(10):704–14.
320. Jovanović I, Živkovic´ M, Jovanović J, et al. The co-inertia approach in identification of specific microRNA in early and advanced atherosclerosis plaque. Med Hypotheses. 2014;83:11–5.
321. Macha MA, Seshacharyulu P, Krishn SR, et al. MicroRNAs (miRNAs) as biomarker(s) for prognosis and diagnosis of gastrointestinal (GI) cancers. Curr Pharm Des. 2014;20(33):5287–97.
322. Williams AH, Valdez G, Moresi V, et al. MicroRNA-206 delays ALS progression and promotes regeneration of neuromuscular synapses in mice. Science. 2009;326:1549–54.
323. Hudson MB, Rahnert JA, Zheng B, et al. miR-182 attenuates atrophy-related gene expression by targeting FoxO3 in skeletal muscle. Am J Physiol Cell Physiol. 2014;307:C314–9.
324. Vickers KC, Landstreet SR, Levin MG, et al. MicroRNA-223 coordinates cholesterol homeostasis. Proc Natl Acad Sci U S A. 2014;111(40):14518–23.
325. Kumar S, Kim CW, Simmons RD, Jo H. Role of flow-sensitive microRNAs in endothelial dysfunction and atherosclerosis—"Mechanosensitive Athero-miRs". Arterioscler Thromb Vasc Biol. 2014;34(10):2206–16.

Lifestyle Choices, Psychological Stress and Their Impact on Ageing: The Role of Telomeres

8

Sergio Davinelli and Immaculata De Vivo

8.1 Introduction

The ends of the chromosomes in all eukaryotic species have specialized, non-coding DNA sequences that, together with associated proteins, are known as telomeres [1]. Telomeric DNA comprises simple tandem repeats of guanine-rich sequences, which are characterized by the hexanucleotide repeat $d(TTAGGG)_n$ in vertebrates. The extreme 3' end of eukaryotic telomeric DNA is single stranded and is typically 100–200 bases long [2, 3]. At birth, human telomeres are typically 10–15 kilobases in length, with substantial inter-individual heterogeneity. On average, human telomeres lose 50–100 base pairs per mitotic division, thus limiting the cell's replicative capacity [4]. The limited replicative capacity of normal cells, initially described in the 1960s by Leonard Hayflick and thus known as the 'Hayflick limit', could be explained based on progressive shortening of telomeres observed with every mitotic event [5, 6]. The telomere length and cell function can be preserved by the reverse transcriptase telomerase. Human telomerase consists of two subunits: an RNA template (TERC, telomerase RNA component) and the catalytic subunit (hTERT, human telomerase reverse transcriptase), which synthesizes the new telomeric DNA from the RNA template [7]. The rate of telomere loss is modifiable by factors other than the mitotic replication rate. Particularly, the GGG triplet within the human telomere sequence TTAGGG is vulnerable to chemical modification. Due to the

S. Davinelli (✉)
Department of Epidemiology, Harvard T. H. Chan School of Public Health,
Boston, MA, USA

Department of Medicine and Health Sciences "V. Tiberio", University of Molise,
Campobasso, Italy
e-mail: sdavinelli@hsph.harvard.edu

I. De Vivo
Department of Epidemiology, Harvard T. H. Chan School of Public Health,
Boston, MA, USA
e-mail: nhidv@channing.harvard.edu

© Springer Nature Switzerland AG 2019
C. Caruso (ed.), *Centenarians*, https://doi.org/10.1007/978-3-030-20762-5_8

content and long stretches of repetitive DNA, it is also thought that telomere sequences suffer disproportionately higher rates of damage by oxidative stress than non-telomeric sequences. Single-stranded breaks of telomeric DNA, caused either directly by reactive oxygen species or indirectly as part of the DNA repair process, are not as efficiently repaired. In addition, telomeric DNA appears to be particularly sensitive to accumulate 8-hydroxydeoxyguanosine, which is one of the major products of DNA oxidation. The shelterin complex, formed by six telomere-specific proteins, shapes and protects chromosome ends. Experimental evidence suggests that oxidative damage at telomeres displaces shelterin proteins, which might be another mechanism by which oxidative stress leads to telomere dysfunction. Telomere shortening may be further increased by chronic low-grade inflammation that affects the ageing process [8, 9]. Importantly, the overexpression of inflammatory mediators associated with telomere-driven cellular senescence can limit the tissue regenerative capacity, compromise the function of tissue-specific stem and progenitor cells and accelerate ageing [9, 10]. Senescent cells have been shown to have a distinct secretome profile, known as the senescence-associated secretory phenotype (SASP). Cells during this state have the ability to produce cytokines, chemokines, growth factors and proteases. Furthermore, telomeric DNA damage can lead to activation of a DNA damage response, which enhances nuclear factor (NF)-κB transcriptional activity. NF-κB activation is responsible for the SASP and can induce (and be activated by) reactive oxygen species generation. Systemic exposures that increase levels of oxidative stress and inflammation (e.g. smoking, obesity and chronic stress) have been associated with shorter telomere lengths in white blood cells [11–13]. On the other hand, it has been hypothesized that healthy lifestyle choices (e.g. lower body mass index and physical activity), tobacco abstinence, a diet high in fruits and vegetables and meditation promote a more stable telomere length, presumably through enhanced antioxidant and anti-inflammatory capability [14, 15] (Fig. 8.1). This chapter summarizes the current knowledge of the

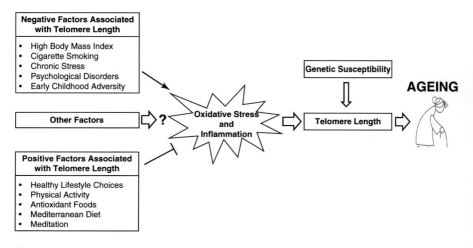

Fig. 8.1 Environmental factors associated with telomere length maintenance

association between telomere length and healthy lifestyle choices and suggests that these factors may play a role in telomere biology maintenance, thus impacting overall health status and longevity.

8.2 Associations Between Lifestyle Choices and Telomere Length Biology

Several studies have observed that a healthy lifestyle is correlated with longer telomeres, likely reflecting protection against age-related diseases. Shorter telomeres are associated with decreased life expectancy and increased rates of developing age-related chronic diseases. Telomere length is deeply influenced by environmental factors and stressful conditions throughout an individual's lifetime. Prospective longitudinal studies have shown that environmental stress exposures perturb telomere homeostasis by increasing oxidative stress and inflammation [16]. Given that multiple environmental factors affect these processes, the objective of the studies reviewed below was to determine the potential relation between environmental/behavioural aspects and telomere length and thus to examine the importance of lifestyle choices in maintaining telomeric stability.

8.2.1 Physical Activity

Regular physical activity has been associated with decreased levels of oxidative stress and inflammation; it also helps to prevent chronic diseases [17, 18]. Evidence from population-based studies on the relation between physical activity and telomere length has been limited and inconsistent, and the potential role of moderate- or vigorous-intensity activity, as well as specific types of physical activity, has remained unclear [19–21]. Moreover, previous studies have not evaluated the association between sedentary behaviour and telomere length. Obese individuals may possess shorter telomeres, and as sedentary lifestyle predicts obesity, sedentary behaviours may influence telomere length [22, 23]. The influence of physical activity, intensity and type of physical activity and sedentary behaviour on telomere length was examined in 7813 women of the Nurses' Health Study [24]. In this large cross-sectional analysis, women who were moderately or highly physically active were found to have longer telomere length than less active women, after adjustment for age and other confounders. This relation remained after additional adjustment for body mass index. Although the association was modest, the difference in telomere length corresponded on average to 4.4 years of ageing, comparable to the difference observed when comparing smokers with non-smokers (4.6 years) [23]. Similarly, greater intensity physical activity (i.e. moderate or vigorous) was associated with longer telomeres independent of body mass index. The longest telomeres were found among women who engaged in moderate or vigorous activities 2–4 hours per week, an amount corresponding to current US guidelines (2.5 h/week) [25]. There was no additional increase in telomere length for most active women compared with those who were moderately active, which suggests that even moderate amounts of activity may influence telomere length. These results extend the literature by showing, in the largest study to date, that even moderate activity, of either moderate or vigorous intensity, may be associated with longer telomeres in middle-aged and older women. Exercise helps to maintain energy balance and to reduce obesity, which may decrease levels of oxidative stress and inflammation [26–28]. Although slightly attenuated, associations remained significant after adjustment for body mass index, suggesting that the relation between activity and longer telomere length may in part be mediated through factors other than body mass. One possibility is that moderate amounts of regular activity may generate low levels of reactive oxygen species that induce adaptive increases in endogenous antioxidant defences, while high amounts of activity may generate excess reactive oxygen species that counteract these defences [29–32]. Although current evidence for this hypothesis is inconsistent, our results and those of others may be consistent with this mechanism [33–35]. Activity may also help to prevent insulin resistance, which has been associated with increased inflammation, oxidative stress and telomere attrition [36–39]. Additionally, activity may help to reduce some negative effects of chronic stress, such as telomere shortening [12, 40]. In summary, physical activity, even in moderate amounts, may be associated with longer telomere length in middle-aged and older women, providing insight into how regular exercise may benefit health on the cellular level.

8.2.2   Diet

Although the influence of food on telomere biology remains under investigation, several epidemiologic studies and randomized clinical trials have investigated the relationship between telomere length and nutrients, foods and dietary patterns. Several food compounds such as vitamins, minerals, omega-3 polyunsaturated fatty acids, and polyphenols have been shown to reduce oxidative stress and chronic inflammation thus affecting telomere length [41].

8.2.2.1 Food Compounds

Multivitamin supplements contain large amounts of many vitamins and minerals and therefore represent a major source of micronutrient intake. Multivitamin use was associated with longer telomeres among 586 women within the Sister Study, a prospective cohort of healthy sisters (age 35–74 years) of breast cancer patients. In general, the use of multivitamin supplements was associated with longer telomere length. Compared with non-users, daily users had on average 5.1% longer telomeres. This difference (273 base pairs) corresponds to $\approx$9.8 years of age-related telomere loss, since each year of age was associated with a 28-base pair decrease in telomere length [42]. Dietary long-chain polyunsaturated fatty acids display anti-inflammatory properties; consequently, their intake is suggested to protect against telomere attrition. Importantly, a balance between omega-3 and omega-6 is necessary to regulate the synthesis of inflammatory mediators and reduce inflammation. Indeed, a randomized controlled 4-month trial showed that it is not omega-3 itself that is important, but rather the ratio between omega-3 and omega-6 fatty acids, since telomere length increases with decreasing omega-3/omega-6 plasma ratios [43]. Another group of dietary components that possess antioxidant and anti-inflammatory properties are polyphenols. Certain polyphenols are found in all plant products (fruit, vegetables, cereals, fruit juices, tea and wine), whereas others are specific to particular foods [44, 45]. Preliminary population studies have observed that these phytochemicals may positively affect telomere length. Green tea and black tea are a rich source of polyphenols such as catechin and epicatechin. Elderly Chinese men who are habitual tea drinkers have longer telomeres than their counterparts who do not drink tea frequently [46]. Another important naturally occurring polyphenol related to telomere length is resveratrol. This compound delays senescence and increases telomere length and telomerase activity but does not extend lifespan in a rodent model [47]. Also, mice that were fed with diets containing curcumin, the active polyphenolic ingredient of the spice turmeric, showed decreased DNA damage and longer telomeres than controls [48]. Additionally, phytochemicals belonging to different chemical classes than polyphenols, particularly carotenoids, have shown positive effects on telomere length. For example, in a population of 786 older (mean age 66 years) individuals from Australia, higher plasma lutein, zeaxanthin and vitamin C concentration was directly correlated to longer telomere [49]. Consistent with these findings, a later study reported a significant direct correlation between longer telomere and dietary intake of β-carotene [50].

8.2.2.2 Mediterranean Diet

The Mediterranean diet has been widely reported to be a model of healthy eating. Meta-analyses of the available prospective cohort studies on the association between adherence to a Mediterranean diet and the onset of age-associated diseases have observed that a greater adherence to a Mediterranean diet is associated with a lower risk of all-cause mortality [51, 52]. Given that fruits, vegetables and nuts—key components of the Mediterranean diet—have well-known antioxidant and anti-inflammatory effects and that telomere length is affected by both these processes, it follows that adherence to the Mediterranean diet would likely be associated with longer telomere length. Preliminary epidemiological studies have recently focussed their attention on the link between the Mediterranean diet and telomere length and its health impact on the general population [53, 54]. For example, this question was addressed in a population of US women within the Nurses' Health Study cohort, an ongoing prospective cohort study of 121,700 nurses enrolled in 1976. Specifically, 4676 disease-free women from nested case-control studies within the NHS, who also completed food frequency questionnaires, were assessed. For comparison, other existing dietary patterns (prudent pattern, Western pattern and Alternative Healthy Eating Index) were also evaluated [55]. After adjustment for potential confounders, greater adherence to the Mediterranean diet by study subjects was associated with longer telomeres. Using a diet score ranging from 0 to 9 points, each 1-point change in diet score corresponded on average to 1.5 years of ageing and

3-point change to 4.5 years of ageing. These results further support the benefits of adherence to the Mediterranean diet for promoting health and longevity.

8.2.3 Psychological Stress

An emerging literature implicates psychological distress and mood disorders, both highly prevalent in women, as potential paths towards accelerated ageing [56, 57]. Prior work identified a relationship between depression and higher levels of inflammatory mediators and oxidative stress [58, 59]. Although less is known about its link to these mechanistic paths, anxiety could also be a risk factor for accelerated morbidity or mortality in ageing. Previous studies have shown that phobic anxiety was significantly related to higher levels of inflammatory markers and an elevated risk of sudden cardiac death and fatal coronary disease [60, 61].

8.2.3.1 Phobic Anxiety
Phobic anxiety is treatable; thus, any potential impact on telomere shortening may be prevented through early identification and treatment. In a study of 5243 women in the Nurses' Health Study, high phobic anxiety was significantly associated with shorter telomere length; this association was consistent after adjustment for confounders such as paternal age-at-birth, smoking, body mass index and physical activity [62]. Although the literature is at an early stage, there is biologic plausibility to support a connection between anxiety and shorter telomeres, particularly through oxidative stress and inflammation. For example, in a study of 362 healthy adults, higher tension-anxiety symptom level was correlated with 8-hydroxydeoxyguanosine, an oxidative DNA damage marker, [63]. In a previous Nurses' Health Study, elevated inflammatory markers (tumour necrosis factor-α receptor II, soluble E-selectin and soluble intercellular adhesion molecule) were observed among diabetic women with the highest phobic anxiety [60]. Similarly, higher scores on the Spielberger State-Trait Anxiety Inventory, used to assess levels of state anxiety and trait anxiety, were significantly correlated with elevated C-reactive protein, interleukin-6 and fibrinogen levels in 853 middle-aged adults [64]. Phobic anxiety is usually the primary presenting condition and is often comorbid with other mental disorders (e.g. depression and substance abuse) [65]. Early intervention may not only mitigate detrimental impact on ageing but could also avert further consequences of accelerated telomere shortening due to secondary development of other mental disorders or serious chronic medical conditions.

8.2.3.2 Psychological Stress and Early Adversity
Exposure to adversity and stress has consistently been associated with a range of negative health outcomes including psychological disorders, immunologic disorders and cardiovascular disease (CVD). Emerging evidence indicates that early childhood may represent a particularly vulnerable time period, as the brain is undergoing rapid neurodevelopmental changes. Although the link between adversity and a range of negative outcomes is established, the mechanism of how these early

experiences alter biological processes has yet to be fully elucidated. Biomarkers of adversity, such as alterations in the hypothalamic-pituitary-adrenal axis and the autonomic nervous system, have been identified and offer insight into this process at a systems level [66, 67]. Telomere shortening may represent an additional cellular level biomarker of adversity. Most recently, psychological stress including a history of early maltreatment, mood disorders, self-reported psychological stress and stress exposure has been associated with shorter telomere length [68–71]. Several studies suggest that acceleration of the cellular ageing process occurs with psychological distress, and this may represent one mechanism by which early adversity is translated into increased morbidity and mortality across health indices. In a study of adults with and without anxiety disorders, across both case and control subjects, childhood adversity was significantly associated with shorter relative telomere length [68]. However, no previous studies have examined whether this association can be demonstrated in children. Early childhood likely represents a critical period for the interaction between stress, cellular ageing and neurodevelopment for several reasons. First, brain development is rapid over the first years of life. Further, early childhood is both a period of rapid telomere attrition and the putative time point at which an individual's rate of telomere length attrition is established epigenetically [72]. Clarification of a temporal relation between adversity exposure and cellular-level biological changes would represent a significant advancement for the study of early life stress. Children living in institutions represent a well-studied model of early adversity. These children receive little attention to their individual needs, are exposed to low-quality caregiving and have limited opportunities to form selective interpersonal attachments. The detrimental impact of institutionalization across biological, social, emotional, neurological and cognitive domains has been established in multiple studies over the last 50 years [73, 74]. In the Bucharest Early Intervention Project, the only longitudinal randomized controlled trial of foster care (compared with continued institutional care) ever conducted, children were recruited from one of six institutions in Romania and randomly assigned to either the 'care as usual' group or the 'foster care' group arm of the clinical intervention. Children who had spent a greater percentage of their early life in institutional care would have significantly shorter telomere length than children who had less exposure to institutional care. A significant inverse association was detected between the percentage of time in institutional care and telomere length in middle childhood. A greater percentage of time in the institution at baseline and at 54 months of age was associated with shorter telomere length in middle childhood. Romanian children with increased early institutional care had shorter telomeres [75]. These findings support data that early childhood experiences including abuse, adversity and serious illness are associated with shorter relative telomere length in adults.

8.2.4 Meditation

Lifestyle behaviours that mitigate the effects of stress may be associated with longer telomere length. Previous research suggests a link between behaviours that focus on the well-being of others, for example, volunteering and caregiving, and overall health and longevity. Forgiveness of others has also been associated with greater longevity. Although the mechanism for this is unclear, the effects may relate to reducing low levels of hostility, which is related to medical morbidity and mortality [76, 77]. Kindness or Metta meditation (Metta from the Pali language of the Buddhist scriptures) is a type of meditation practice that focusses on developing a positive intention, unselfish kindness and warmth towards all people. Because shorter telomeres are associated with chronic psychological stress and loving-kindness meditation appears to decrease stress, telomere length was examined in a population of meditators. Moreover, it was hypothesized that these subjects would have longer telomeres than those of age-, gender-, and education-matched controls. The meditators had longer telomere length than controls. Especially among women, loving-kindness meditation practitioners had significantly longer telomere length than controls, which remained significant even after controlling for body mass index and past depression. Furthermore, in gender-stratified analysis, women meditators had significantly longer telomeres than gender-matched controls. The association between loving-kindness meditation and relative telomere length was stronger in women than in men, but the reason for this is not clear. One explanation may be that the women who were experienced meditators spent more time practicing meditation [78]. The finding of longer telomeres in meditators is consistent with previous work that demonstrated an association between meditation and telomerase [79] and suggests the possibility that meditation could have beneficial effects on telomere length, a marker of cellular ageing linked to longevity.

8.3   Telomeres and Mortality Risk

Although the association between telomere length and mortality remains unclear, there is rapidly growing literature exploring whether there is an overall association between telomere length and subsequent mortality risk. In population-based prospective studies, it has repeatedly been shown that individuals with short telomeres have an increased risk for cardiovascular events, stroke and all-cause mortality [80–82]. Although several studies reported an association between short telomeres and cause-specific mortality or all-cause mortality, there is a substantial variability among the findings of these studies. A number of factors may contribute to the variation observed in the relationship between telomere length and mortality due to the different measurement techniques and the varying age, sex and diet or ethnicity of the study participants. Regarding cause-specific mortality and the association between CVD and telomere length, numerous studies indicate a modest inverse link [83, 84] but are heterogeneous with regard to CVD mortality [85, 86]. The associations of telomere length with cancer are probably more complex. Although many cancer cells express highly active telomerase, their telomeres are shorter than those in normal tissue. Indeed, multiple studies have found increased risk of cancer incidence with short telomere length [87–89]. However, more recent studies investigating cancer-specific associations have also correlated longer telomere length with increased risk of several tumours [90–92]. With regard to all-cause mortality, a study by Cawthon et al. reported for the first time that telomere shortening contributed to all-cause mortality in 143 US subjects aged 60–97 years [93]. Rode et al. conducted the largest study so far ($n = 64{,}637$) to demonstrate that short telomeres were associated with a higher risk of all-cause mortality [94]. More recently, a large meta-analysis was conducted to evaluate the association of telomere length with all-cause mortality, taking advantage of both previously published results from cohort studies of the general population and unpublished original data from the Swedish Twin Registry. Interestingly, the magnitude of the association of telomere length and all-cause mortality was similar for the youngest groups (<75 years and 75–80 years) but weaker for the oldest old (over 80 years). Therefore, it seems that the oldest old have lower hazard of telomere attrition-associated all-cause mortality increment than younger individuals.

8.4   Conclusion

Telomeres are repetitive DNA sequences at the ends of eukaryotic chromosomes that undergo attrition each time a somatic cell divides. The capping function of telomeres protects the physical integrity of chromosomes and prevents the loss of genomic DNA. Telomere length is an important factor in the pathobiology of human disease. Oxidative stress and inflammation are important modulators of telomere loss, and it has been shown that telomere attrition is accelerated by these processes. Environmental factors may perturb telomere length homeostasis by increasing the levels of oxidative stress and inflammation. Therefore, shorter telomeres may represent a marker of the cumulative burden of inflammation and oxidative stress. Individuals who lead a healthy lifestyle by increasing their physical activity, practicing meditation, adhering to the Mediterranean diet and using multivitamins have been shown to have longer telomeres than those who do not adhere to such lifestyle behaviours. The studies reviewed here highlight the influence of lifestyle factors on telomere biology. Strategic management of these factors may have important implications for ageing and longevity. Finally, the associations between telomere length, age-related disease and mortality should inspire further research to unravel the factors contributing to the individual differences in telomere length.

References

1. Stewart SA, Weinberg RA. Telomeres: cancer to human aging. Annu Rev Cell Dev Biol. 2006;22:531–57.
2. McElligott R, Wellinger RJ. The terminal DNA structure of mammalian chromosomes. EMBO J. 1997;16(12):3705–14.
3. Huffman KE, Levene SD, Tesmer VM, Shay JW, Wright WE. Telomere shortening is proportional to the size of the G-rich telomeric 3′-overhang. J Biol Chem. 2000;275(26):19719–22.
4. Palm W, de Lange T. How shelterin protects mammalian telomeres. Annu Rev Genet. 2008;42:301–34.

5. Harley CB, Futcher AB, Greider CW. Telomeres shorten during ageing of human fibroblasts. Nature. 1990;345(6274):458–60.
6. Hayflick L, Moorhead PS. The serial cultivation of human diploid cell strains. Exp Cell Res. 1961;25:585–621.
7. Xin H, Liu D, Songyang Z. The telosome/shelterin complex and its functions. Genome Biol. 2008;9(9):232.
8. Barnes RP, Fouquerel E, Opresko PL. The impact of oxidative DNA damage and stress on telomere homeostasis. Mech Ageing Dev. 2019;177:37–45. pii: S0047-6374(18)30052-6.
9. Jurk D, Wilson C, Passos JF, Oakley F, Correia-Melo C, Greaves L, et al. Chronic inflammation induces telomere dysfunction and accelerates ageing in mice. Nat Commun. 2014;2: 4172.
10. Sahin E, Depinho RA. Linking functional decline of telomeres, mitochondria and stem cells during ageing. Nature. 2010;464(7288):520–8.
11. Butt HZ, Atturu G, London NJ, Sayers RD, Bown MJ. Telomere length dynamics in vascular disease: a review. Eur J Vasc Endovasc Surg. 2010;40(1):17–26.
12. Epel ES, Blackburn EH, Lin J, Dhabhar FS, Adler NE, Morrow JD, et al. Accelerated telomere shortening in response to life stress. Proc Natl Acad Sci U S A. 2004;101(49):17312–5.
13. Simon NM, Smoller JW, McNamara KL, Maser RS, Zalta AK, Pollack MH, et al. Telomere shortening and mood disorders: preliminary support for a chronic stress model of accelerated aging. Biol Psychiatry. 2006;60(5):432–5.
14. Cherkas LF, Hunkin JL, Kato BS, Richards JB, Gardner JP, Surdulescu GL, et al. The association between physical activity in leisure time and leukocyte telomere length. Arch Intern Med. 2008;168(2):154–8.
15. Mirabello L, Huang WY, Wong JY, Chatterjee N, Reding D, Crawford ED, et al. The association between leukocyte telomere length and cigarette smoking, dietary and physical variables, and risk of prostate cancer. Aging Cell. 2009;8(4):405–13.
16. Starkweather AR, Alhaeeri AA, Montpetit A, Brumelle J, Filler K, Montpetit M, et al. An integrative review of factors associated with telomere length and implications for biobehavioral research. Nurs Res. 2014;63(1):36–50.
17. Ji LL, Gomez-Cabrera MC, Vina J. Exercise and hormesis: activation of cellular antioxidant signaling pathway. Ann N Y Acad Sci. 2006;1067:425–35.
18. Kasapis C, Thompson PD. The effects of physical activity on serum C-reactive protein and inflammatory markers: a systematic review. J Am Coll Cardiol. 2005;45(10):1563–9.
19. Ludlow AT, Zimmerman JB, Witkowski S, Hearn JW, Hatfield BD, Roth SM. Relationship between physical activity level, telomere length, and telomerase activity. Med Sci Sports Exerc. 2008;40(10):1764–71.
20. Woo J, Tang N, Leung J. No association between physical activity and telomere length in an elderly Chinese population 65 years and older. Arch Intern Med. 2008;168(19):2163–4.
21. Zhu H, Wang X, Gutin B, Davis CL, Keeton D, Thomas J, et al. Leukocyte telomere length in healthy Caucasian and African-American adolescents: relationships with race, sex, adiposity, adipokines, and physical activity. J Pediatr. 2011;158(2):215–20.
22. Hu FB, Li TY, Colditz GA, Willett WC, Manson JE. Television watching and other sedentary behaviors in relation to risk of obesity and type 2 diabetes mellitus in women. JAMA. 2003;289(14):1785–91.
23. Valdes AM, Andrew T, Gardner JP, Kimura M, Oelsner E, Cherkas LF, et al. Obesity, cigarette smoking, and telomere length in women. Lancet. 2005;366(9486):662–4.
24. Du M, Prescott J, Kraft P, Han J, Giovannucci E, Hankinson SE, et al. Physical activity, sedentary behavior, and leukocyte telomere length in women. Am J Epidemiol. 2012;175(5): 414–22.
25. US Department of Health and Human Services. Physical activity guidelines for Americans: be active, healthy, and happy! Washington, DC: US Department of Health and Human Services; 2008.
26. McTiernan A. Mechanisms linking physical activity with cancer. Nat Rev Cancer. 2008;8(3):205–11.

27. Keaney JF Jr, Larson MG, Vasan RS, Wilson PW, Lipinska I, Corey D, et al. Obesity and systemic oxidative stress: clinical correlates of oxidative stress in the Framingham study. Arterioscler Thromb Vasc Biol. 2003;23(3):434–9.

28. Pou KM, Massaro JM, Hoffmann U, Vasan RS, Maurovich-Horvat P, Larson MG, et al. Visceral and subcutaneous adipose tissue volumes are cross-sectionally related to markers of inflammation and oxidative stress: the Framingham heart study. Circulation. 2007;116(11):1234–41.

29. Covas MI, Elosua R, Fitó M, Alcántara M, Coca L, Marrugat J. Relationship between physical activity and oxidative stress biomarkers in women. Med Sci Sports Exerc. 2002;34(5):814–9.

30. Poulsen HE, Loft S, Vistisen K. Extreme exercise and oxidative DNA modification. J Sports Sci. 1996;14(4):343–6.

31. Powers SK, Jackson MJ. Exercise-induced oxidative stress: cellular mechanisms and impact on muscle force production. Physiol Rev. 2008;88(4):1243–76.

32. Radak Z, Chung HY, Goto S. Systemic adaptation to oxidative challenge induced by regular exercise. Free Radic Biol Med. 2008;44(2):153–9.

33. Collins M, Renault V, Grobler LA, St Clair Gibson A, Lambert MI, et al. Athletes with exercise-associated fatigue have abnormally short muscle DNA telomeres. Med Sci Sports Exerc. 2003;35(9):1524–8.

34. Magi F, Dimauro I, Margheritini F, Duranti G, Mercatelli N, Fantini C, et al. Telomere length is independently associated with age, oxidative biomarkers, and sport training in skeletal muscle of healthy adult males. Free Radic Res. 2018;52(6):639–47.

35. Rae DE, Vignaud A, Butler-Browne GS, Thornell LE, Sinclair-Smith C, Derman EW, et al. Skeletal muscle telomere length in healthy, experienced, endurance runners. Eur J Appl Physiol. 2010;109(2):323–30.

36. Al-Attas OS, Al-Daghri NM, Alokail MS, Alfadda A, Bamakhramah A, Sabico S, et al. Adiposity and insulin resistance correlate with telomere length in middle-aged Arabs: the influence of circulating adiponectin. Eur J Endocrinol. 2010;163(4):601–7.

37. Ceriello A, Motz E. Is oxidative stress the pathogenic mechanism underlying insulin resistance, diabetes, and cardiovascular disease? The common soil hypothesis revisited. Arterioscler Thromb Vasc Biol. 2004;24(5):816–23.

38. Demissie S, Levy D, Benjamin EJ, Cupples LA, Gardner JP, Herbert A, et al. Insulin resistance, oxidative stress, hypertension, and leukocyte telomere length in men from the Framingham heart study. Aging Cell. 2006;5(4):325–30.

39. Gardner JP, Li S, Srinivasan SR, Chen W, Kimura M, Lu X, et al. Rise in insulin resistance is associated with escalated telomere attrition. Circulation. 2005;111(17):2171–7.

40. Puterman E, Lin J, Blackburn E, O'Donovan A, Adler N, Epel E. The power of exercise: buffering the effect of chronic stress on telomere length. PLoS One. 2010;5(5):e10837.

41. Rafie N, Golpour Hamedani S, Barak F, Safavi SM, Miraghajani M. Dietary patterns, food groups and telomere length: a systematic review of current studies. Eur J Clin Nutr. 2017;71(2):151–8.

42. Xu Q, Parks CG, DeRoo LA, Cawthon RM, Sandler DP, Chen H. Multivitamin use and telomere length in women. Am J Clin Nutr. 2009;89(6):1857–63.

43. Kiecolt-Glaser JK, Epel ES, Belury MA, Andridge R, Lin J, Glaser R, et al. Omega-3 fatty acids, oxidative stress, and leukocyte telomere length: a randomized controlled trial. Brain Behav Immun. 2013;28:16–24.

44. Davinelli S, Maes M, Corbi G, Zarrelli A, Willcox DC, Scapagnini G. Dietary phytochemicals and neuro-inflammaging: from mechanistic insights to translational challenges. Immun Ageing. 2016;13:16.

45. Davinelli S, Scapagnini G. Polyphenols: a promising nutritional approach to prevent or reduce the progression of prehypertension. High Blood Press Cardiovasc Prev. 2016;23(3):197–202.

46. Chan R, Woo J, Suen E, Leung J, Tang N. Chinese tea consumption is associated with longer telomere length in elderly Chinese men. Br J Nutr. 2010;103(1):107–13.

47. da Luz PL, Tanaka L, Brum PC, Dourado PM, Favarato D, Krieger JE, et al. Red wine and equivalent oral pharmacological doses of resveratrol delay vascular aging but do not extend life span in rats. Atherosclerosis. 2012;224(1):136–42.

48. Thomas P, Wang YJ, Zhong JH, Kosaraju S, O'Callaghan NJ, Zhou XF, et al. Grape seed polyphenols and curcumin reduce genomic instability events in a transgenic mouse model for Alzheimer's disease. Mutat Res. 2009;661(1–2):25–34.
49. Sen A, Marsche G, Freudenberger P, Schallert M, Toeglhofer AM, Nagl C, et al. Association between higher plasma lutein, zeaxanthin, and vitamin C concentrations and longer telomere length: results of the Austrian stroke prevention study. J Am Geriatr Soc. 2014;62(2): 222–9.
50. Marcon F, Siniscalchi E, Crebelli R, Saieva C, Sera F, Fortini P, et al. Diet-related telomere shortening and chromosome stability. Mutagenesis. 2012;27(1):49–57.
51. Sofi F, Cesari F, Abbate R, Gensini GF, Casini A. Adherence to Mediterranean diet and health status: meta-analysis. BMJ. 2008;337:a1344.
52. Schwingshackl L, Hoffmann G. Adherence to Mediterranean diet and risk of cancer: a systematic review and meta-analysis of observational studies. Int J Cancer. 2014;135(8):1884–97.
53. Boccardi V, Esposito A, Rizzo MR, Marfella R, Barbieri M, Paolisso G. Mediterranean diet, telomere maintenance and health status among elderly. PLoS One. 2013;8(4):e62781.
54. Davinelli S, Trichopoulou A, Corbi G, De Vivo I, Scapagnini G. The potential nutrigeroprotective role of Mediterranean diet and its functional components on telomere length dynamics. Ageing Res Rev. 2019;49:1–10. https://doi.org/10.1016/j.arr.2018.11.001.
55. Crous-Bou M, Fung TT, Prescott J, Julin B, Du M, Sun Q, et al. Mediterranean diet and telomere length in Nurses' health study: population based cohort study. BMJ. 2014;349:g6674.
56. Wolkowitz OM, Epel ES, Reus VI, Mellon SH. Depression gets old fast: do stress and depression accelerate cell aging? Depress Anxiety. 2010;27(4):327–38.
57. Epel ES. Psychological and metabolic stress: a recipe for accelerated cellular aging? Hormones (Athens). 2009;8(1):7–22.
58. Maes M. The cytokine hypothesis of depression: inflammation, oxidative & nitrosative stress (IO&NS) and leaky gut as new targets for adjunctive treatments in depression. Neuro Endocrinol Lett. 2008;29(3):287–91.
59. Scapagnini G, Davinelli S, Drago F, De Lorenzo A, Oriani G. Antioxidants as antidepressants: fact or fiction? CNS Drugs. 2012;26(6):477–90.
60. Brennan AM, Fargnoli JL, Williams CJ, Li T, Willett W, Kawachi I, et al. Phobic anxiety is associated with higher serum concentrations of adipokines and cytokines in women with diabetes. Diabetes Care. 2009;32(5):926–31.
61. Albert CM, Chae CU, Rexrode KM, Manson JE, Kawachi I. Phobic anxiety and risk of coronary heart disease and sudden cardiac death among women. Circulation. 2005;111(4):480–7.
62. Okereke OI, Prescott J, Wong JY, Han J, Rexrode KM, De Vivo I. High phobic anxiety is related to lower leukocyte telomere length in women. PLoS One. 2012;7(7):e40516.
63. Irie M, Asami S, Nagata S, Ikeda M, Miyata M, Kasai H. Psychosocial factors as a potential trigger of oxidative DNA damage in human leukocytes. Jpn J Cancer Res. 2001;92(3):367–76.
64. Pitsavos C, Panagiotakos DB, Papageorgiou C, Tsetsekou E, Soldatos C, Stefanadis C. Anxiety in relation to inflammation and coagulation markers, among healthy adults: the ATTICA study. Atherosclerosis. 2006;185(2):320–6.
65. Kessler RC, Ruscio AM, Shear K, Wittchen HU. Epidemiology of anxiety disorders. Curr Top Behav Neurosci. 2010;2:21–35.
66. Evans GW. A multimethodological analysis of cumulative risk and allostatic load among rural children. Dev Psychol. 2003;39(5):924–33.
67. Shonkoff JP, Boyce WT, McEwen BS. Neuroscience, molecular biology, and the childhood roots of health disparities: building a new framework for health promotion and disease prevention. JAMA. 2009;301(21):2252–9.
68. Kananen L, Surakka I, Pirkola S, Suvisaari J, Lönnqvist J, Peltonen L, et al. Childhood adversities are associated with shorter telomere length at adult age both in individuals with an anxiety disorder and controls. PLoS One. 2010;5(5):e10826.
69. Lung FW, Chen NC, Shu BC. Genetic pathway of major depressive disorder in shortening telomeric length. Psychiatr Genet. 2007;17(3):195–9.

70. Parks CG, Miller DB, McCanlies EC, Cawthon RM, Andrew ME, DeRoo LA, et al. Telomere length, current perceived stress, and urinary stress hormones in women. Cancer Epidemiol Biomark Prev. 2009;18(2):551–60.
71. Tyrka AR, Price LH, Kao HT, Porton B, Marsella SA, Carpenter LL. Childhood maltreatment and telomere shortening: preliminary support for an effect of early stress on cellular aging. Biol Psychiatry. 2010;67(6):531–4.
72. Cameron N, Demerath EW. Critical periods in human growth and their relationship to diseases of aging. Am J Phys Anthropol. 2002;Suppl 35:159–84.
73. Nelson CA 3rd, Zeanah CH, Fox NA, Marshall PJ, Smyke AT, Guthrie D. Cognitive recovery in socially deprived young children: the Bucharest Early Intervention Project. Science. 2007;318(5858):1937–40.
74. Pollak SD, Nelson CA, Schlaak MF, Roeber BJ, Wewerka SS, Wiik KL, et al. Neurodevelopmental effects of early deprivation in postinstitutionalized children. Child Dev. 2010;81(1):224–36.
75. Drury SS, Theall K, Gleason MM, Smyke AT, De Vivo I, Wong JY, et al. Telomere length and early severe social deprivation: linking early adversity and cellular aging. Mol Psychiatry. 2012;17(7):719–27.
76. Toussaint LL, Owen AD, Cheadle A. Forgive to live: forgiveness, health, and longevity. J Behav Med. 2012;35(4):375–86.
77. Smith TW, Glazer K, Ruiz JM, Gallo LC. Hostility, anger, aggressiveness, and coronary heart disease: an interpersonal perspective on personality, emotion, and health. J Pers. 2004;72(6):1217–70.
78. Hoge EA, Chen MM, Orr E, Metcalf CA, Fischer LE, Pollack MH, et al. Loving-kindness meditation practice associated with longer telomeres in women. Brain Behav Immun. 2013;32:159–63.
79. Jacobs TL, Epel ES, Lin J, Blackburn EH, Wolkowitz OM, Bridwell DA, et al. Intensive meditation training, immune cell telomerase activity, and psychological mediators. Psychoneuroendocrinology. 2011;36(5):664–81.
80. Willeit P, Willeit J, Brandstätter A, Ehrlenbach S, Mayr A, Gasperi A, et al. Cellular aging reflected by leukocyte telomere length predicts advanced atherosclerosis and cardiovascular disease risk. Arterioscler Thromb Vasc Biol. 2010;30(8):1649–56.
81. Brouilette SW, Moore JS, McMahon AD, Thompson JR, Ford I, Shepherd J, et al. Telomere length, risk of coronary heart disease, and statin treatment in the West of Scotland primary prevention study: a nested case-control study. Lancet. 2007;369(9556):107–14.
82. Needham BL, Rehkopf D, Adler N, Gregorich S, Lin J, Blackburn EH, et al. Leukocyte telomere length and mortality in the National Health and Nutrition Examination Survey, 1999–2002. Epidemiology. 2015;26(4):528–35.
83. Farzaneh-Far R, Cawthon RM, Na B, Browner WS, Schiller NB, Whooley MA. Prognostic value of leukocyte telomere length in patients with stable coronary artery disease: data from the heart and soul study. Arterioscler Thromb Vasc Biol. 2003;28(7):1379–84.
84. Fitzpatrick AL, Kronmal RA, Gardner JP, Psaty BM, Jenny NS, Tracy RP, et al. Leukocyte telomere length and cardiovascular disease in the cardiovascular health study. Am J Epidemiol. 2007;165(1):14–21.
85. Svensson J, Karlsson MK, Ljunggren Ö, Tivesten Å, Mellström D, Movérare-Skrtic S. Leukocyte telomere length is not associated with mortality in older men. Exp Gerontol. 2014;57:6–12.
86. Fitzpatrick AL, Kronmal RA, Kimura M, Gardner JP, Psaty BM, Jenny NS, et al. Leukocyte telomere length and mortality in the cardiovascular health study. J Gerontol A Biol Sci Med Sci. 2011;66(4):421–9.
87. Ma H, Zhou Z, Wei S, Liu Z, Pooley KA, Dunning AM, et al. Shortened telomere length is associated with increased risk of cancer: a meta-analysis. PLoS One. 2011;6(6):e20466.
88. Prescott J, Wentzensen IM, Savage SA, De Vivo I. Epidemiologic evidence for a role of telomere dysfunction in cancer etiology. Mutat Res. 2012;730(1–2):75–84.

89. Wentzensen IM, Mirabello L, Pfeiffer RM, Savage SA. The association of telomere length and cancer: a meta-analysis. Cancer Epidemiol Biomark Prev. 2011;20(6):1238–50.
90. Xie H, Wu X, Wang S, Chang D, et al. Long telomeres in peripheral blood leukocytes are associated with an increased risk of soft tissue sarcoma. Cancer. 2013;119(10):1885–91.
91. Sanchez-Espiridion B, Chen M, Chang JY, Lu C, Chang DW, Roth JA, et al. Telomere length in peripheral blood leukocytes and lung cancer risk: a large case-control study in Caucasians. Cancer Res. 2014;74(9):2476–86.
92. Han J, Qureshi AA, Prescott J, Guo Q, Ye L, Hunter DJ, et al. A prospective study of telomere length and the risk of skin cancer. J Invest Dermatol. 2009;129(2):415–21.
93. Cawthon RM, Smith KR, O'Brien E, Sivatchenko A, Kerber RA. Association between telomere length in blood and mortality in people aged 60 years or older. Lancet. 2003;361(9355):393–5.
94. Rode L, Nordestgaard BG, Bojesen SE. Peripheral blood leukocyte telomere length and mortality among 64,637 individuals from the general population. J Natl Cancer Inst. 2015;107(6):djv074.

Gut Microbiota Pattern of Centenarians

9

Lu Wu, Angelo Zinellu, Luciano Milanesi, Salvatore Rubino, David J. Kelvin, and Ciriaco Carru

9.1 Introduction

As the global population becomes an ageing population, to prolong a healthy lifespan in the elderly is rapidly becoming a vital challenge for modern medical research. Centenarians, an extremely aged population with extended healthy lifespan, have been used as an ideal model to study ageing and longevity [1–4]. The rapid growth of the centenarian population around the world is a challenge and a burden for the society. Thus, extra concerns are also needed to improve centenarian quality of life.

Gut microbiota, dominated by bacteria, not only facilitates the digestion of food and the generation of metabolites essential for various functions but also mediates the effects of exogenous chemicals that can modulate host activities by gut–brain cross-talk [5, 6]. For instance, in the gut, a process modulated by microbiota is responsible for most of the production of 5-hydroxytryptamine, which can

L. Wu
Division of Immunology, International Institute of Infection and Immunity, Shantou University Medical College, Shantou, Guangdong, China

Department of Biomedical Sciences, University of Sassari, Sassari, Italy
e-mail: wulu@sibs.ac.cn

A. Zinellu · S. Rubino · C. Carru (✉)
Department of Biomedical Sciences, University of Sassari, Sassari, Italy
e-mail: azinellu@uniss.it; rubino@uniss.it; carru@uniss.it

L. Milanesi
Institute of Biomedical Technologies, National Research Council of Italy, Segrate, Italy
e-mail: luciano.milanesi@itb.cnr.it

D. J. Kelvin (✉)
Division of Immunology, International Institute of Infection and Immunity, Shantou University Medical College, Shantou, Guangdong, China

Department of Microbiology and Immunology, Dalhousie University, Halifax, NS, Canada
e-mail: dkelvin@jidc.org

© Springer Nature Switzerland AG 2019
C. Caruso (ed.), *Centenarians*, https://doi.org/10.1007/978-3-030-20762-5_9

comprehensively affect host locomotor activation and addictive behaviour [7]. Gut microbiota also participates in host immunity and response to infections [8, 9]. Gut microbiota shows high diversity and plasticity in response to different environmental factors such as diet, lifestyle and drugs [10–14]. Throughout life, the gut microbiota has co-evolved with the host to adapt; this includes changes in nutrition derived from the host diet and immune response sensitivity.

Ageing is characterized by physiological, genomic and metabolic changes, and the immune function also declines with ageing [15, 16]. Ageing affects the gut microbiota as well [12, 17, 18]. In the elderly, the gut microbiota is closely related to host health status [19, 20]. Intriguingly, as a highly dynamic and relative stable ecosystem, gut microbiota can be rapidly altered by diet [10, 21, 22]. Gut microbiota not only can be comparatively easily manipulated, such as by diet intervention and calorie restriction [23, 24], but also can shape the function of the host immune system and exert systemic metabolic effects [25, 26]. Thus, gut microbiota might be a contributor for the human longevity and a promising target for diagnostics of the health status. Since gut microbiota, composition and function can be modulated, it has great promise in promoting healthy ageing.

Many clinical issues accompanying ageing, such as bowel disorders, cardiovascular diseases, constipation and Parkinson's disease, are also closely correlated to perturbations in the composition and function of gut microbiota [27–32]. Interestingly, centenarians, as a successful extreme ageing model, can delay or escape these age-related diseases; thus, it is critical to gain insight into the gut microbiota in centenarians and explore the possible contribution of the gut microbiota to ageing and longevity in humans.

In this chapter, we review the possible associations between gut microbiota and longevity in humans. We also discuss how the gut microbiota, integrating with the host immunity and metabolism, contributes to longevity. In addition, we summarize the features of gut microbiota in longevity across populations.

9.2   Network of Longevity and Gut Microbiota in Humans

As discussed in Chap. 1, longevity of centenarians is a complex biological phenotype, and different factors contribute to this phenotype [24, 33–39]. The longevity has also been demonstrated to be associated with defined metabolites such as primary bile acids, short-chain fatty acids (SCFA) and isocitrate, closely related to gut microbiota activities [24, 40, 41]. The human gut microbiota is largely determined by environmental factors such as diet and nutrition, but it is also influenced by the host genetic factors [10, 11, 13, 14, 42]. Gut microbiota has a fundamental function in shaping and training the immunity system [9, 43], and it is also involved in age-related inflammation [44]. In addition, the gut microbiota-derived metabolites such as SCFA, trimethylamine (TMA), secondary bile acids, cholate and indoles can modulate the host metabolism and regulate the host immunity and gene expression [32, 45–51]. For example, primary bile acids are produced in the liver from cholesterol and metabolized in the intestine by the gut microbiota into secondary bile acids. The microbial modifications of bile acids influence host metabolism through

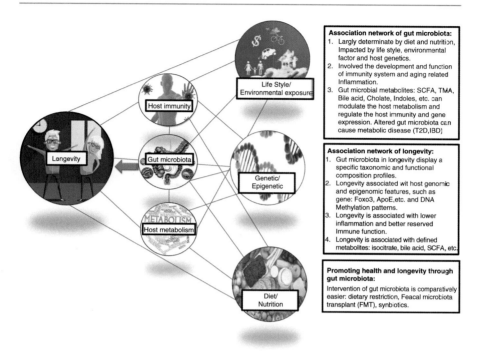

Association network of gut microbiota:
1. Largely determinate by diet and nutrition, Impacted by life style, environmental factor and host genetics.
2. Involved the development and function of immunity system and aging related Inflammation.
3. Gut microbial metabolites: SCFA, TMA, Bile acid, Cholate, Indoles, etc. can modulate the host metabolism and regulate the host immunity and gene expression. Altered gut microbiota can cause metabolic disease (T2D,IBD)

Association network of longevity:
1. Gut microbiota in longevity display a specific taxonomic and functional composition profiles.
2. Longevity associated wit host genomic and epigenomic features, such as gene: Foxo3, ApoE,etc. and DNA Methylation patterns.
3. Longevity is associated with lower inflammation and better reserved Immune function.
4. Longevity is associated with defined metabolites: isocitrate, bile acid, SCFA, etc.

Promoting health and longevity through gut microbiota:
Intervention of gut microbiota is comparatively easier: dietary restriction, Feacal microbiota transplant (FMT), synbiotics.

Life Style/ Environmental exposure

Host immunity

Longevity

Gut microbiota

Genetic/ Epigenetic

Host metabolism

Diet/ Nutrition

Fig. 9.1 The association networks of longevity in humans. For explanation, see text

the nuclear farnesoid X receptor and the G-protein-coupled membrane receptor 5 [49]. Conversely, bile acids can modulate gut microbial composition both directly and indirectly through the activation of innate immune genes in the small intestine [52]. Besides, previous studies have already identified specific gut microbiota features in longevity populations [41, 53–56]. Given the complex interaction between human gut microbiota, immunity, metabolism and longevity, the gut microbiota interacts with dietary, lifestyle and other environmental factors, as well as with the genetic and epigenetic factors of the host. Furthermore, gut microbiota affects other longevity contributors such as host metabolism and immunity and forms a complex network of contributors to longevity, and it could be regarded as a putative potential contributor to human longevity. We summarize the association networks of the gut microbiota and longevity in Fig. 9.1.

In animal models, such as *Caenorhabditis elegans* and turquoise killifish (*Nothobranchius furzeri*), which have a simple gut microbiota and a short and easily monitored lifespan, gut microbiota has been proved to influence the host longevity [57, 58]. In human beings, the gut microbiota is a highly complicated, dynamic and personalized ecosystem [59]. Although the longevity features of gut microbiota have been demonstrated, due to the difficulties of longitudinal studies to trace the ageing process in humans, identifying the mechanistic contribution of gut microbiota to lifespan in humans is intrinsically challenging. There is still no direct evidence of causal relationship between gut microbiota and longevity in humans. However, as discussed below, a series of breakthrough studies have pointed out the possible association between gut microbiota and human longevity.

9.3 Gut Microbiota, Host Metabolism and Longevity

The association between the distinct metabolomic signatures and human longevity has been identified [40, 41, 60]. Human gut microbiota might contribute to longevity by interacting with the host metabolism. For instance, a longitudinal study in a big cohort from the USA had found that high concentrations of the citric acid cycle intermediate isocitrate and the bile acid taurocholate are negatively associated with longevity [60]. Another cross-sectional study in China has found that the centenarians showed a distinct metabolic pattern. Seven characteristic components closely related to the centenarians were identified, namely acetic acid, total SCFA, Mn, Co, propionic acid, butyric acid and valeric acid. Their concentrations were significantly higher in the centenarian group [24]. These longevity-related metabolites are also closely related to the gut microbiota. Take SCFAs as an example. SCFAs are the principal products of colonic fermentation of dietary fibres by the anaerobic intestinal microbiota. Absorbed by the colonocytes in the gut, they provide around 10% of the daily caloric requirements in humans [61, 62]. SCFAs also inhibit stimuli-induced expression of adhesion molecules and chemokine production and consequently suppress monocyte/macrophage and neutrophil recruitment, suggesting an anti-inflammatory action [51]. Moreover, SCFAs are critical modulators of epigenetic changes and stimulate the epithelial cells to release molecules to facilitate the brain–gut cross-talk [47]. Other gut microbial metabolites such as TMA, ethylphenyl sulphate, indole and propionic acid are also involved in the host health and disease [8, 63].

9.4 Gut Microbiota, Host Immunity and Longevity

Ageing is accompanied by the decline in physiological functions and the increase in the incidence of infectious diseases and chronic inflammatory systemic diseases [64]. Diseases of inflammatory origin, such as atherosclerosis, cardiovascular disease, type II diabetes, arthritis, dementia, Alzheimer's disease, osteoporosis and cancer, are widespread in the elderly, delayed or escaped in centenarians [65, 66]. Although previous studies have found that centenarians cannot avoid the alteration of the immune system observed in aged people, such as the decrease of CD3+ T lymphocytes, helper CD4+ T lymphocytes, cytotoxic T CD8+ lymphocytes and CD19+ B lymphocytes [36], some immune parameters are better preserved in centenarians than in the elderly. It can be observed by an increase in the number of cells with markers of NK activity and T lymphocytes able to mediate non-MHC-restricted cytotoxicity [36, 67] (Chap. 3). Moreover, low levels of interleukin (IL)-6 (pro-inflammatory cytokine) and high levels of IL-10, IL1-Ra and TGF-β1 (anti-inflammatory) are associated with lower inflammatory status in centenarians [37, 39] (Chap. 1). Therefore, the specific characteristics of the immune system in centenarians suggest its potential contribution to host longevity and healthy ageing.

The host immune system and the gut microbiota have complex interactions [68]. As previously stated, gut microbiota plays a fundamental role in the induction, training and function of the host immune system [9, 43]. The dysbiosis of the gut microbiota leads to a decrease in the resistance to the colonization of pathogens and to an increase in pathological immune responses, hence contributing to the inflammatory status [44]. Studies in mice have revealed that the ageing-related inflammation was mostly induced by the ageing-associated gut microbiota. Thus, to reverse these age-related microbiota changes represents a potential strategy for reducing age-associated inflammation [69, 70]. In humans, evidence has also shown that the ageing process deeply affected the structure of the human gut microbiota [12, 53, 54], as well as its homeostasis with the host immune system [25, 55]. Due to its crucial role in the host immunity, it is possible that the gut microbiota contributes to human longevity by its involvement in modulating the host immunity status.

9.5   The Features of Gut Microbiota in Longevity

Throughout life, the gut microbiota undergoes a co-evolution with the human host, adapting itself to the progressive changes of the host gut environment. Several studies, performed in different cohorts, have compared the gut microbiota of centenarians with that of young adults and the elderly (Table 9.1) [53–56, 71–73].

The overall structure of gut microbiota in centenarians was evaluated based on α- and β-diversity when compared within each cohort. For instance, α-diversity is an indicator of the complexity of the gut microbial ecosystem for each individual. Low α-diversity of the gut microbiota is associated with fragile and disturbed gut ecosystems [12, 74]. In centenarians, the gut microbiota α-diversity was reported to be higher than that observed in the control elderly [53, 54] although not in all the studies [55, 71]. This indicates that the functional metabolic potential of intestinal microbiota in centenarians is diverse from that observed in controls. β-Diversity is used to measure the differentiation of gut microbiota among individuals in a "group" scale. In centenarians, the gut microbiota β-diversity was reported to be higher than that observed in the control elderly [53, 56], although not in all the studies [54, 55]. The wide variation within centenarians might be a consequence of different adaptation of gut microbiota to ageing. Overall, the diverse features of gut microbiota across populations may be caused by non-standardized recruitment strategies, different genetic background, different diets, different environmental factors, different methodologies of processing the samples and data.

Among the observed gut microbiota features in centenarians (Table 9.2), some seem universal, such as the lower abundance of *Faecalibacterium*, a dominant genus in gut microbiota, which is widely reported to decline with ageing [20]. Other features are unique to defined population; for example, the enrichment of *Methanobrevibacter* in centenarians was detected in a Sardinian (Italy) centenarian

Table 9.1 Summary of centenarian gut microbiota studies

Ref.	Cohort location	Recruited subjects number (age range)			Sequencing strategies		α diversity (taxonomic)		α diversity (functional)		β diversity (taxonomic) Variation among individuals within longevity groups	β diversity (functional)
		Young	Elderly	Long-living	Sequencing platform	Targeted region	Shannon index	Richness	Shannon index	Richness		
a	Sardinia (Italy)	19 (21–33)	23 (68–88)	19 (99–107)	Illumina Hiseq and Miseq	Shotgun metagenomic sequencing and 16S rRNA V3V4	≈	≈	↑	↑	↑	↑
[53]	Bologna (Italy)	20 (25–40)	43 (59–78)	21 (99–104)	Phylogentic microarry and qPCR	16S rRNA	→	NA	NA	NA	→	NA
[55]	Bologna (Italy)	15 (22–48)	15 (65–75)	39 (99–109)	Illumina Miseq	16S rRNA V3V4	↑	↑	NA	NA	↑	NA
[54]	Sichuan (China)	47 (24–64)	54 (65–83)	67 (90–102)	Illumina Miseq	16S rRNA V4V5	↑	↑	NA	NA	≈	NA
[56]	Bama (China)		16 (80–99)	8 (100–108)	Illumina Miseq	16S rRNA V4	≈	↑	NA	NA	↑	NA
[71]	Japanese	187 (20–59)	91 (60–89)	25 (90–100)	Illumina Miseq	16S rRNA V3V4	→	→	NA	NA		NA

The Shannon diversity index is commonly used to characterize species diversity in a community. It accounts for both abundance and evenness of the species present [http://www.tiem.utk.edu/~gross/bioed/bealsmodules/shannonDI.html]
α-diversity is the mean species diversity in sites or habitats at a local scale. β-diversity refers to the differentiation among those habitats. [https://en.wikipedia.org/wiki/Alpha_diversity#cite_note-Whittaker1960-1]
[a]Author unpublished observations

Table 9.2 Summary of the features of gut microbiota in centenarians compared with control groups within each cohort according to different studies

Genus/species that differently distributed	a	[53]	[55]	[54]	[56]	[71]
g_Bifidobacterium	↑	↓	↑			
g_Bilophila	↑		↑			
g_Escherichia	↑	↑			↑	↑
g_Blautia	↓	↓		↑		
g_Akkermansia	≈	≈	↑		↓	
g_Butyricimonas	↑		↑	↑	↓	
g_Christensenella	↑		↑	↑		
g_Clostridium	↓					
g_Lactobacillus					↓	
g_Coprococcus	↓	↓	↓	↑	↓	↓
g_Desulfovibrio	↑					↑
g_Dorea	↓					
g_Faecalibacterium	↓	↓	↓	↓	↓	↓
g_Leuconostoc	↑					
g_Methanobrevibacter	↑		↑			
g_Odoribacter			↑		↓	
g_Oscillibacter	↑					
g_Parabacteroides	↑			↑	↓	
g_Pseudomonas		↑		↑		
g_Pyramidobacter	↑					
g_Roseburia	↓	↓	↓		↑	↓
g_Ruminococcus	↓					

[a]Author unpublished observations

cohort but not in other centenarian cohorts [53, 54, 56]. The enrichment of *Bifidobacterium* was also observed in centenarians from Sardinia and in the semi-supercentenarians (>105 years) from Emilia-Romagna (Italy) but not in other cohorts [53]. *Akkermansia* was found in low abundance in centenarians from Sichuan, China, and Sardinia but enriched in the semi-supercentenarians and not in the centenarians (100–105 years) from Emilia-Romagna. Overall, independent studies on intestinal microbiota in centenarians reveal that the gut microbiota in longevity populations has accumulated several sub-dominant species.

The structural variation of the taxonomic composition of gut microbiota in centenarians corresponds to the metabolic functional differentiation of gut microbiota in centenarians compared with that in the elderly. For instance, the proportion of the two most abundant phyla in the gut, the *Firmicutes/Bacteroidetes* proportion (F/B), is an important index of the structure of gut microbiota. It has already been found to be associated with body mass index, obesity, production of SCFA and ageing [75–79]. A high F/B ratio might be associated with an increased energy harvest [80]. The F/B ratio in Sardinian and Emilia-Romagnan centenarians showed an interesting trait: lower than that observed in the elderly group but similar to that observed in young people [55]. The low ratio of F/B in centenarians may lead to the decline of energy harvesting capabilities in centenarians. Centenarians in Sardinia were found to have lower gene pathway abundance involved in complex carbohydrate degradation, which correlates with the significantly lower prevalence of *Ruminococcus* and *Faecalibacterium* in the gut than that of the young and elderly. Furthermore, in centenarians, the enrichment of gene pathways related to the utilization of energy by microbes through glycolysis correlated with the enrichment of *Enterococcus, Lactobacillus* and *Escherichia* [81]. Diminished physical activities and energy expenditure are associated with the ageing process and may cause the decline of energy requirements for humans [82, 83]. Therefore, overall, the gut microbiota in centenarians has the potential to decrease the energy harvest by the host, which seems an adaptation to ageing.

9.6   Conclusion

As a significant contributor to human health, gut microbiota may play a critical role during the ageing process. Using centenarians as an extreme ageing and longevity model, studies have identified the centenarian-specific gut microbiota characters. To establish the causal relationship between the gut microbiota and longevity, further studies based on the animal models are needed to interpret the role of gut microbiota in longevity.

Funding LKSF (DJK) and International Institute of Infection and Immunity, Shantou University Medical College (DJK) and Dalhousie Medical Research Foundation (DJK); Shantou University Medical College and University of Sassari Joint Ph.D. program (LW). DJK is a recipient of a Tier I Canada Research Chair in Vaccinology and Inflammation; "Ministero dell'Università e della Ricerca" Ministero dell'Istruzione, dell'Università e della Ricerca (MIUR, Italy)—Progetti di Ricerca di Rilevante Interesse Nazionale—PRIN 2015 (Prot. 20157ATSLF_002) (CC, AZ) and MIUR Consiglio Nazionale delle Ricerche Flagship InterOmics (cod. PB05) (CC, AZ and LM).

The funders had no role in study design, data collection and analysis, decision to publish, or preparation of the manuscript.

References

1. Brooks-Wilson AR. Genetics of healthy aging and longevity. Hum Genet. 2013;132(12):1323–38.
2. Caselli G, Pozzi L, Vaupel JW, Deiana L, Pes G, Carru C, et al. Family clustering in Sardinian longevity: a genealogical approach. Exp Gerontol. 2006;41(8):727–36.
3. Santoro A, Ostan R, Candela M, Biagi E, Brigidi P, Capri M, et al. Gut microbiota changes in the extreme decades of human life: a focus on centenarians. Cell Mol Life Sci. 2018;75(1): 129–48.
4. Santos-Lozano A, Santamarina A, Pareja-Galeano H, Sanchis-Gomar F, Fiuza-Luces C, Cristi-Montero C, et al. The genetics of exceptional longevity: insights from centenarians. Maturitas. 2016;90:49–57.
5. Clemmensen C, Muller TD, Woods SC, Berthoud HR, Seeley RJ, Tschop MH. Gut-brain cross-talk in metabolic control. Cell. 2017;168(5):758–74.
6. Lee WJ, Hase K. Gut microbiota-generated metabolites in animal health and disease. Nat Chem Biol. 2014;10(6):416–24.

7. Filip M, Bader M. Overview on 5-HT receptors and their role in physiology and pathology of the central nervous system. Pharmacol Rep. 2009;61(5):761–77.
8. Sonnenburg JL, Backhed F. Diet-microbiota interactions as moderators of human metabolism. Nature. 2016;535(7610):56–64.
9. Thaiss CA, Zmora N, Levy M, Elinav E. The microbiome and innate immunity. Nature. 2016;535(7610):65–74.
10. David LA, Maurice CF, Carmody RN, Gootenberg DB, Button JE, Wolfe BE, et al. Diet rapidly and reproducibly alters the human gut microbiome. Nature. 2014;505(7484):559–63.
11. Goodrich JK, Waters JL, Poole AC, Sutter JL, Koren O, Blekhman R, et al. Human genetics shape the gut microbiome. Cell. 2014;159(4):789–99.
12. O'Toole PW, Jeffery IB. Gut microbiota and aging. Science. 2015;350(6265):1214–5.
13. Rampelli S, Schnorr SL, Consolandi C, Turroni S, Severgnini M, Peano C, et al. Metagenome sequencing of the Hadza hunter-gatherer gut microbiota. Curr Biol. 2015;25(13):1682–93.
14. Xie H, Guo R, Zhong H, Feng Q, Lan Z, Qin B, et al. Shotgun metagenomics of 250 adult twins reveals genetic and environmental impacts on the gut microbiome. Cell Syst. 2016;3(6):572–84.e3.
15. Lopez-Otin C, Blasco MA, Partridge L, Serrano M, Kroemer G. The hallmarks of aging. Cell. 2013;153(6):1194–217.
16. Montecino-Rodriguez E, Berent-Maoz B, Dorshkind K. Causes, consequences, and reversal of immune system aging. J Clin Invest. 2013;123(3):958–65.
17. Maffei VJ, Kim S, Blanchard ET, Luo M, Jazwinski SM, Taylor CM, et al. Biological aging and the human gut microbiota. J Gerontol A Biol Sci Med Sci. 2017;72(11):1474–82.
18. An R, Wilms E, Masclee AAM, Smidt H, Zoetendal EG, Jonkers D. Age-dependent changes in GI physiology and microbiota: time to reconsider? Gut. 2018;67(12):2213–22.
19. O'Toole PW, Jeffery IB. Microbiome-health interactions in older people. Cell Mol Life Sci. 2018;75(1):119–28.
20. Jeffery IB, Lynch DB, O'Toole PW. Composition and temporal stability of the gut microbiota in older persons. ISME J. 2016;10(1):170–82.
21. Bashan A, Gibson TE, Friedman J, Carey VJ, Weiss ST, Hohmann EL, et al. Universality of human microbial dynamics. Nature. 2016;534(7606):259–62.
22. Faith JJ, Guruge JL, Charbonneau M, Subramanian S, Seedorf H, Goodman AL, et al. The long-term stability of the human gut microbiota. Science. 2013;341(6141):1237439.
23. Fontana L, Partridge L. Promoting health and longevity through diet: from model organisms to humans. Cell. 2015;161(1):106–18.
24. Cai D, Zhao S, Li D, Chang F, Tian X, Huang G, et al. Nutrient intake is associated with longevity characterization by metabolites and element profiles of healthy centenarians. Nutrients. 2016;8(9).
25. Claesson MJ, Jeffery IB, Conde S, Power SE, O'Connor EM, Cusack S, et al. Gut microbiota composition correlates with diet and health in the elderly. Nature. 2012;488(7410):178–84.
26. Ottaviani E, Ventura N, Mandrioli M, Candela M, Franchini A, Franceschi C. Gut microbiota as a candidate for lifespan extension: an ecological/evolutionary perspective targeted on living organisms as metaorganisms. Biogerontology. 2011;12(6):599–609.
27. Hill-Burns EM, Debelius JW, Morton JT, Wissemann WT, Lewis MR, Wallen ZD, et al. Parkinson's disease and Parkinson's disease medications have distinct signatures of the gut microbiome. Mov Disord. 2017;32(5):739–49.
28. Heintz-Buschart A, Pandey U, Wicke T, Sixel-Doring F, Janzen A, Sittig-Wiegand E, et al. The nasal and gut microbiome in Parkinson's disease and idiopathic rapid eye movement sleep behavior disorder. Mov Disord. 2018;33(1):88–98.
29. Manichanh C, Borruel N, Casellas F, Guarner F. The gut microbiota in IBD. Nat Rev Gastroenterol Hepatol. 2012;9(10):599–608.
30. Norman JM, Handley SA, Baldridge MT, Droit L, Liu CY, Keller BC, et al. Disease-specific alterations in the enteric virome in inflammatory bowel disease. Cell. 2015;160(3):447–60.
31. Li J, Zhao F, Wang Y, Chen J, Tao J, Tian G, et al. Gut microbiota dysbiosis contributes to the development of hypertension. Microbiome. 2017;5(1):14.

32. Zhu W, Gregory JC, Org E, Buffa JA, Gupta N, Wang Z, et al. Gut microbial metabolite TMAO enhances platelet hyperreactivity and thrombosis risk. Cell. 2016;165(1):111–24.
33. Beekman M, Blanche H, Perola M, Hervonen A, Bezrukov V, Sikora E, et al. Genome-wide linkage analysis for human longevity: genetics of healthy aging study. Aging Cell. 2013;12(2):184–93.
34. Poulain M, Pes GM, Grasland C, Carru C, Ferrucci L, Baggio G, et al. Identification of a geographic area characterized by extreme longevity in the Sardinia island: the AKEA study. Exp Gerontol. 2004;39(9):1423–9.
35. Gentilini D, Mari D, Castaldi D, Remondini D, Ogliari G, Ostan R, et al. Role of epigenetics in human aging and longevity: genome-wide DNA methylation profile in centenarians and centenarians' offspring. Age. 2013;35(5):1961–73.
36. Sansoni P, Cossarizza A, Brianti V, Fagnoni F, Snelli G, Monti D, et al. Lymphocyte subsets and natural killer cell activity in healthy old people and centenarians. Blood. 1993;82(9):2767–73.
37. Lio D, Scola L, Crivello A, Colonna-Romano G, Candore G, Bonafe M, et al. Inflammation, genetics, and longevity: further studies on the protective effects in men of IL-10-1082 promoter SNP and its interaction with TNF-alpha-308 promoter SNP. J Med Genet. 2003;40(4):296–9.
38. Sansoni P, Vescovini R, Fagnoni F, Biasini C, Zanni F, Zanlari L, et al. The immune system in extreme longevity. Exp Gerontol. 2008;43(2):61–5.
39. Bonafe M, Olivieri F, Cavallone L, Giovagnetti S, Mayegiani F, Cardelli M, et al. A gender-dependent genetic predisposition to produce high levels of IL-6 is detrimental for longevity. Eur J Immunol. 2001;31(8):2357–61.
40. Montoliu I, Scherer M, Beguelin F, DaSilva L, Mari D, Salvioli S, et al. Serum profiling of healthy aging identifies phospho- and sphingolipid species as markers of human longevity. Aging (Albany NY). 2014;6(1):9–25.
41. Collino S, Montoliu I, Martin FP, Scherer M, Mari D, Salvioli S, et al. Metabolic signatures of extreme longevity in northern Italian centenarians reveal a complex remodeling of lipids, amino acids, and gut microbiota metabolism. PLoS One. 2013;8(3):e56564.
42. Rothschild D, Weissbrod O, Barkan E, Kurilshikov A, Korem T, Zeevi D, et al. Environment dominates over host genetics in shaping human gut microbiota. Nature. 2018;555(7695):210–5.
43. Honda K, Littman DR. The microbiota in adaptive immune homeostasis and disease. Nature. 2016;535(7610):75–84.
44. Buford TW. (Dis)Trust your gut: the gut microbiome in age-related inflammation, health, and disease. Microbiome. 2017;5(1):80.
45. Rowland I, Gibson G, Heinken A, Scott K, Swann J, Thiele I, et al. Gut microbiota functions: metabolism of nutrients and other food components. Eur J Nutr. 2018;57(1):1–24.
46. van de Wouw M, Schellekens H, Dinan TG, Cryan JF. Microbiota-gut-brain axis: modulator of host metabolism and appetite. J Nutr. 2017;147(5):727–45.
47. Bhat MI, Kapila R. Dietary metabolites derived from gut microbiota: critical modulators of epigenetic changes in mammals. Nutr Rev. 2017;75(5):374–89.
48. Lin R, Liu W, Piao M, Zhu H. A review of the relationship between the gut microbiota and amino acid metabolism. Amino Acids. 2017;49(12):2083–90.
49. Wahlstrom A, Sayin SI, Marschall HU, Backhed F. Intestinal crosstalk between bile acids and microbiota and its impact on host metabolism. Cell Metab. 2016;24(1):41–50.
50. den Besten G, van Eunen K, Groen AK, Venema K, Reijngoud DJ, Bakker BM. The role of short-chain fatty acids in the interplay between diet, gut microbiota, and host energy metabolism. J Lipid Res. 2013;54(9):2325–40.
51. Vinolo MA, Rodrigues HG, Nachbar RT, Curi R. Regulation of inflammation by short chain fatty acids. Nutrients. 2011;3(10):858–76.
52. Inagaki T, Moschetta A, Lee YK, Peng L, Zhao G, Downes M, et al. Regulation of antibacterial defense in the small intestine by the nuclear bile acid receptor. Proc Natl Acad Sci U S A. 2006;103(10):3920–5.
53. Biagi E, Franceschi C, Rampelli S, Severgnini M, Ostan R, Turroni S, et al. Gut microbiota and extreme longevity. Curr Biol. 2016;26(11):1480–5.

54. Kong F, Hua Y, Zeng B, Ning R, Li Y, Zhao J. Gut microbiota signatures of longevity. Curr Biol. 2016;26(18):R832–3.
55. Biagi E, Nylund L, Candela M, Ostan R, Bucci L, Pini E, et al. Through ageing, and beyond: gut microbiota and inflammatory status in seniors and centenarians. PLoS One. 2010;5(5):e10667.
56. Wang F, Yu T, Huang G, Cai D, Liang X, Su H, et al. Gut microbiota community and its assembly associated with age and diet in Chinese centenarians. J Microbiol Biotechnol. 2015;25(8):1195–204.
57. Smith P, Willemsen D, Popkes M, Metge F, Gandiwa E, Reichard M, et al. Regulation of life span by the gut microbiota in the short-lived African turquoise killifish. eLife 2017;6.
58. Han B, Sivaramakrishnan P, Lin CJ, Neve IAA, He J, Tay LWR, et al. Microbial genetic composition tunes host longevity. Cell. 2017;169(7):1249–62.e13.
59. Caporaso JG, Lauber CL, Costello EK, Berg-Lyons D, Gonzalez A, Stombaugh J, et al. Moving pictures of the human microbiome. Genome Biol. 2011;12(5):R50.
60. Cheng S, Larson MG, McCabe EL, Murabito JM, Rhee EP, Ho JE, et al. Distinct metabolomic signatures are associated with longevity in humans. Nat Commun. 2015;6:6791.
61. Binder HJ. Role of colonic short-chain fatty acid transport in diarrhea. Annu Rev Physiol. 2010;72:297–313.
62. Bergman EN. Energy contributions of volatile fatty acids from the gastrointestinal tract in various species. Physiol Rev. 1990;70(2):567–90.
63. Wang Z, Zhao Y. Gut microbiota derived metabolites in cardiovascular health and disease. Protein Cell. 2018;9(5):416–31.
64. Ginaldi L, De Martinis M, D'Ostilio A, Marini L, Loreto MF, Quaglino D. Immunological changes in the elderly. Aging (Milano). 1999;11(5):281–6.
65. Freund A, Orjalo AV, Desprez PY, Campisi J. Inflammatory networks during cellular senescence: causes and consequences. Trends Mol Med. 2010;16(5):238–46.
66. Pedro VC, Arturo RH, Alejandro PM, Oscar RC. Sociodemographic and clinical characteristics of centenarians in Mexico City. Biomed Res Int. 2017;2017:7195801.
67. Thompson JS, Wekstein DR, Rhoades JL, Kirkpatrick C, Brown SA, Roszman T, et al. The immune status of healthy centenarians. J Am Geriatr Soc. 1984;32(4):274–81.
68. Kamada N, Seo SU, Chen GY, Nunez G. Role of the gut microbiota in immunity and inflammatory disease. Nat Rev Immunol. 2013;13(5):321–35.
69. Thevaranjan N, Puchta A, Schulz C, Naidoo A, Szamosi JC, Verschoor CP, et al. Age-associated microbial dysbiosis promotes intestinal permeability, systemic inflammation, and macrophage dysfunction. Cell Host Microbe. 2017;21(4):455–66.e4.
70. Fransen F, van Beek AA, Borghuis T, Aidy SE, Hugenholtz F, van der Gaast-de Jongh C, et al. Aged gut microbiota contributes to systemical inflammaging after transfer to germ-free mice. Front Immunol. 2017;8:1385.
71. Odamaki T, Kato K, Sugahara H, Hashikura N, Takahashi S, Xiao JZ, et al. Age-related changes in gut microbiota composition from newborn to centenarian: a cross-sectional study. BMC Microbiol. 2016;16:90.
72. Rampelli S, Candela M, Turroni S, Biagi E, Collino S, Franceschi C, et al. Functional metagenomic profiling of intestinal microbiome in extreme ageing. Aging (Albany NY). 2013;5(12):902–12.
73. Park SH, Kim KA, Ahn YT, Jeong JJ, Huh CS, Kim DH. Comparative analysis of gut microbiota in elderly people of urbanized towns and longevity villages. BMC Microbiol. 2015;15:49.
74. Qin J, Li Y, Cai Z, Li S, Zhu J, Zhang F, et al. A metagenome-wide association study of gut microbiota in type 2 diabetes. Nature. 2012;490(7418):55–60.
75. Koliada A, Syzenko G, Moseiko V, Budovska L, Puchkov K, Perederiy V, et al. Association between body mass index and firmicutes/bacteroidetes ratio in an adult Ukrainian population. BMC Microbiol. 2017;17(1):120.
76. Fernandes J, Su W, Rahat-Rozenbloom S, Wolever TM, Comelli EM. Adiposity, gut microbiota and faecal short chain fatty acids are linked in adult humans. Nutr Diabetes. 2014;4:e121.
77. Ley RE, Turnbaugh PJ, Klein S, Gordon JI. Microbial ecology: human gut microbes associated with obesity. Nature. 2006;444(7122):1022–3.

78. Turnbaugh PJ, Hamady M, Yatsunenko T, Cantarel BL, Duncan A, Ley RE, et al. A core gut microbiome in obese and lean twins. Nature. 2009;457(7228):480–4.
79. Mariat D, Firmesse O, Levenez F, Guimaraes V, Sokol H, Dore J, et al. The firmicutes/bacteroidetes ratio of the human microbiota changes with age. BMC Microbiol. 2009;9:123.
80. Turnbaugh PJ, Ley RE, Mahowald MA, Magrini V, Mardis ER, Gordon JI. An obesity-associated gut microbiome with increased capacity for energy harvest. Nature. 2006;444(7122):1027–31.
81. Chassard C, Lacroix C. Carbohydrates and the human gut microbiota. Curr Opin Clin Nutr Metab Care. 2013;16(4):453–60.
82. Frisard MI, Broussard A, Davies SS, Roberts LJ 2nd, Rood J, de Jonge L, et al. Aging, resting metabolic rate, and oxidative damage: results from the Louisiana Healthy Aging Study. J Gerontol A Biol Sci Med Sci. 2007;62(7):752–9.
83. Ruggiero C, Metter EJ, Melenovsky V, Cherubini A, Najjar SS, Ble A, et al. High basal metabolic rate is a risk factor for mortality: the Baltimore longitudinal study of aging. J Gerontol A Biol Sci Med Sci. 2008;63(7):698–706.

Impact of Mediterranean Diet on Longevity

10

Antonia Trichopoulou and Vassiliki Benetou

10.1 Introduction

Ageing has emerged as one of the major policy issues of our times, since both the proportion and the absolute number of older adults are constantly rising worldwide [1]. At the same time, it has become increasingly important not only to live longer but also to live in good health. In 2016, the *Global strategy and action plan on ageing and health (2016–2020)* was adopted by the World Health Assembly with the aim to promote healthy ageing and contribute to achieving the vision that all people can live long and healthy lives [2]. Nutrition was identified as one of the key behaviours that influence healthy ageing, not only through the prevention of chronic, non-communicable diseases and age-related conditions, but also through the preservation of functional ability and independence. Thus, elucidating the role of nutrition in the ageing process is considered very important for the promotion of health and wellbeing in the older age.

Convincing evidence has accumulated over the years that specific foods, food groups, as well as dietary patterns are associated with reduced incidence of, and mortality from, major chronic diseases and longer survival [3, 4]. Research interest on the dietary patterns, more specifically the combinations, quantities and frequency with which foods are habitually consumed, has been dictated by the ability of the latter to integrate complex or subtle interactive effects of many dietary exposures, to accommodate the inter-correlation of nutrients within a diet and to bypass problems

---

Dedicated to our dear friend and colleague Christina Bamia

A. Trichopoulou (✉)
Hellenic Health Foundation, Athens, Greece
e-mail: atrichopoulou@hhf-greece.gr

V. Benetou
Department of Hygiene, Epidemiology and Medical Statistics, School of Medicine,
National and Kapodistrian University of Athens, Athens, Greece
e-mail: vbenetou@med.uoa.gr

© Springer Nature Switzerland AG 2019
C. Caruso (ed.), *Centenarians*, https://doi.org/10.1007/978-3-030-20762-5_10

created by multiple testing. Furthermore, studying diet by means of dietary pattern analysis is an appealing method to assess the influence of the overall diet, which most likely possesses synergistic and cumulative effects on health and disease, in contrast to individual foods and nutrients, while it offers a more intuitive understanding of the findings [3, 5, 6]. Dietary patterns can be defined either in a hypothesis-oriented approach (*a priori* methods), based on the available scientific evidence for the relation of specific foods with particular diseases (e.g. Mediterranean diet pattern, healthy eating index) or in an exploratory approach (*a posteriori* methods) that relies on available empirical data without a priori hypothesis [6]. Irrespective of the methodology used, the study of dietary patterns in relation to health is now a fundamental step in the process of formulating food-based dietary guidelines [7, 8].

The *Consortium on Health and Ageing: Network of Cohorts in Europe and the United States* (CHANCES), a collaborative, large-scale project funded by the European Commission and coordinated by Hellenic Health Foundation in Greece, provided a unique opportunity to investigate the association between adherence to *a priori* healthy dietary patterns and important health outcomes among elderly populations. Adherence to the Healthy Diet Indicator (HDI), an index based on World Health Organization (WHO) dietary recommendations (issued in 2003) and all-cause mortality, was tested using harmonized data from 396,391 participants, aged 60 years and older, from 11 prospective cohorts in Europe and the United States. A ten-point increase in the HDI (representing adherence to one additional WHO guideline) was associated with a 10% decrease in total mortality for men and women combined [9]. These estimates translate to an increased life expectancy of 2 years at the age of 60 years and highlight the importance of adhering to a healthy dietary pattern even in the older age. Similar associations were observed for cancer incidence, as greater adherence to the World Cancer Research Fund/American Institute for Cancer Research (WCRF/AICR) dietary recommendations for the prevention of cancer, measured by an appropriate index, was associated with lower risk of cancer among 362,114 older adults from seven prospective cohorts, participants in the CHANCES project [10].

In the next part of this chapter, we will focus on the impact of Mediterranean diet (MD), a well-known *a priori* healthy dietary pattern, on longevity.

10.2   Mediterranean Diet Pattern and Survival

Traditional Mediterranean diet prevailed in the olive tree-growing areas of the Mediterranean region up to the early 1960s. It is characterized by high intake of vegetables, fruits, legumes and cereals (mainly in unprocessed forms); low intake of meat and meat products; low-to-moderate intake of dairy products; moderate-to-high intake of fish; high intake of unsaturated added lipids, particularly in the form of olive oil and modest intake of ethanol, mainly as wine during meals [11].

The beneficial health effects of the traditional Mediterranean diet were first pointed out by the classic studies of Ancel Keys [12]. Although these studies were essentially ecological and they invoked assumptions that were not adequately documented, they did point to an explanation of a paradox: that countries like Greece, with unhealthy lifestyles, including high prevalence of smokers and suboptimal health care systems, had in the 1960s some of the longest adult life expectancies. Almost by exclusion, diet was considered the underlying benefactor for countries like Greece, Spain, Italy and France.

An advance in the study of MD was the introduction of a score by Trichopoulou and colleagues in the 1990s assessing the degree of adherence to the traditional Mediterranean diet [13]. Since then, this score, as well as variants of it proposed by other investigators, have been used in numerous epidemiological studies assessing adherence to the MD in relation to longevity and incidence of and mortality from several major chronic diseases and outcomes [14–22].

With consistency unusual in biomedical research, adherence to MD has been inversely associated with total mortality, incidence of and mortality from cardiovascular diseases and malignant neoplasms. These associations were evident among populations living in different geographical areas, at different points in time and by different investigators using mainly observational [13, 15, 16, 18, 19] but also experimental study designs [17, 20, 21, 23].

In the context of the *European Prospective Investigation Into Cancer and nutrition (EPIC)* study, greater adherence to MD was associated with lower overall cancer risk based on a sample of 142,605 men and 335,873 women from 11 European countries [24]. The reduction in risk was evident in both Mediterranean and non-Mediterranean countries, with a somewhat stronger protection among smokers and against tobacco-related cancers. It was estimated that 4.7% of cancers among men and 2.4% among women would be avoided if study subjects had a greater adherence to MD.

In the context of the *PREDIMED* study, a multicentred randomized trial conducted in Spain, 7447 participants (55–80 years of age, 57% women) at high cardiovascular risk were assigned to one of three diets: an MD supplemented with extra-virgin olive oil, an MD supplemented with mixed nuts or a control diet (with advice to reduce dietary fat) [23]. After a median follow-up of 4.8 years, incidence of major cardiovascular events was lower among those assigned to MD supplemented with extra-virgin olive oil or nuts, than among those assigned to a reduced-fat diet.

Reduced incidence of other major chronic diseases with great burden among older adults, such as type 2 diabetes mellitus, neurodegenerative diseases such as dementia and hip fractures, in relation to higher adherence to MD has also been reported [16, 25–27].

In relation to all-cause mortality, the most recent comprehensive meta-analysis of prospective cohort studies, conducted by Eleftheriou and colleagues, quantified the association of adherence to MD, as well as its components, with all-cause mortality [18]. Based on 30 studies (with 225,600 deaths) published until January 2018, adherence to MD was significantly inversely associated with all-cause mortality. More specifically, when comparing the highest to the lowest adherence category, as well as per one standard deviation increment in MD scale, there was a 21% [summary relative risk (sRR): 0.79; 95% confidence intervals (CIs): 0.77–0.81] and a 8% (sRR: 0.92; 95% CI: 0.90–0.94) decrease in all-cause mortality, respectively. With respect to the individual dietary components of MD that influence mostly this association, relatively stronger and statistically significant inverse associations were highlighted for moderate versus none-to-excessive alcohol consumption (sRR: 0.86; 95% CIs: 0.77–0.97) and for above- versus below-the-median consumptions of fruits (sRR: 0.88, 95% CIs: 0.83–0.94) and vegetables (sRR: 0.94; 95% CI: 0.89–0.98), whereas a positive association was apparent for above- versus below-the-median intake of meat (sRR: 1.07; 95% CI: 1.01–1.13). Another analysis by Trichopoulou and colleagues based on the Greek segment of the EPIC study investigated the relative importance of the individual components of the MD in generating the inverse association of increased adherence to this diet and overall mortality. The main dietary components of the MD which predicted lower total mortality were moderate consumption of ethanol, low consumption of meat and meat products, and high consumption of vegetables, fruits and nuts, olive oil and legumes [28]. On the other hand, it should be noted that the reduction in mortality in relation to the MD score was apparent even though no strong associations with mortality were evident for each of the individual components of the score [13]. Several explanations have been postulated for this consistent observation, such as (a) individual components may have small effects that emerge only when the components are integrated into a simple, unidimensional score and (b) there may be biological interactions between different components of the MD that could be difficult to detect unless very large samples are used.

The plausible biological mechanisms through which MD could exert its beneficial effects to health and promote longevity have been mostly attributed to dietary constituents of the key food groups consumed in moderate-to-high amounts in MD.

Thus, vitamins, phytochemicals, antioxidants (e.g. vit C, E and A), flavonoids, potassium, fibre and folate, minerals and fibre (abundant in plant foods), ethanol and polyphenols in wine (the most frequent type of alcoholic drink consumed in the traditional MD), oleic acid abundant in olive oil and omega-3 fatty acids abundant in fish contribute through anti-inflammatory, antithrombotic, antioxidative and antioncogenic pathways and favourable metabolic effects [3, 29]. Additionally, as mentioned before, nutrient–nutrient interactions and food synergies between the individual components of this pattern consumed as whole foods have shown to exist, possibly conferring an advantageous biological activity.

Recently, the underlying mechanisms that may be involved in the inverse association of MD with cardiovascular disease were investigated in *Women's Health Study* using data from 25,994 women who were followed up for 12 years [22]. Higher baseline MD intake was associated with approximately one-fourth relative risk reduction in CVD events. The potential mediating effects of approximately 40 biomarkers were evaluated, among which, biomarkers of inflammation, glucose metabolism, insulin resistance and adiposity contributed the most in the observed association.

10.3 Mediterranean Diet in the Third Millennium: Just Prevention or Even Therapy?

There is accumulating evidence that MD, beyond its role in primary prevention, could be beneficial for individuals with already diagnosed chronic diseases, offering safe and effective, alternative, non-pharmaceutical therapeutic options that not only could provide symptom relief but also could slow the progression of the disease.

Indeed, MD has been linked to reductions in joint inflammation in patients with rheumatoid arthritis [30]. Furthermore, data from the *Osteoarthritis Initiative* indicate that adherence to MD is associated with better quality of life, decreased pain, disability and depressive symptoms among individuals with knee osteoarthritis or at high risk of developing knee osteoarthritis [31]. A significant improvement in knee flexion and hip rotation for those who adhere to MD was also observed in patients with osteoarthritis [32].

Greater adherence to the traditional MD has been associated with a significant reduction in mortality among individuals diagnosed with coronary heart disease [33]. In the context of the *Lyon Diet Heart Study*, a randomized secondary prevention trial, a Mediterranean-type diet was associated with a reduced rate of recurrence after a first myocardial infarction [34]. Findings from another study support the hypothesis that a "Mediterranean-like" diet is associated with lower coronary artery calcification progression and a lower degree of coronary artery calcification in both men and women [35]. Adopting MD has also been associated with improved glycaemic control and lower cardiovascular risk in persons with established type 2 diabetes [36]. Adherence to MD might also be beneficial among cancer survivors. Higher adherence to MD was associated with better overall survival in long-term colon cancer survivors [37].

Thus, evidence for the therapeutic and prognostic aspects of MD is steadily accumulating, and MD should be studied further for its potential to contribute as an alternative, tasty and pleasant option for secondary prevention at an individual and population level.

10.4 Longevity, Traditional Foods and Sustainable Environment in the Mediterranean Diet

In 2010, the Mediterranean diet was inscribed on UNESCO's Lists of Intangible Cultural Heritage. By then, research on diet and health had already provided convincing evidence that a diet that adheres to the principles of the traditional Mediterranean diet is associated with longer survival. This could be partly attributed to Mediterranean traditional foods on which this diet largely relies. The estimation of the micronutrient content of traditional Greek foods in relation to professional recommendations showed a rich micronutrient profile [38].

As most traditional Mediterranean foods are plant-based and integrated to the local bio-system and economy, they are almost, by definition, environmentally friendly or at least minimally disturbing to the ecosystem. This has been demonstrated in several studies, which have indicated that Mediterranean diet has a lower environmental impact than other dietary patterns [39].

Thus, there is a need to highlight sustainable food production and consumption as interconnected elements, with the purpose of promoting health, as well as preserving the link between local food products, nutrition, food safety, sustainability and biodiversity. Of note, non-Mediterranean countries could also incorporate several components of the traditional Mediterranean diet in their food system, keeping in mind what the traditional Mediterranean diet is and what it is not [40].

Despite being widely documented and acknowledged as a healthy diet, the MD is paradoxically becoming less the diet of choice in most Mediterranean countries. The erosion of the Mediterranean diet heritage is alarming, as it has undesirable impacts not only on health but also on social, cultural, economic and environmental trends in the Mediterranean region.

The broader understanding of the many sustainable benefits of the MD can contribute to its revitalization by increasing its current perception from simply a healthy diet to a sustainable lifestyle model.

10.5   Conclusion

The Mediterranean diet and lifestyle were shaped by climatic conditions, poverty and hardship, rather than by intellectual insight or wisdom. Nevertheless, results from methodologically superior nutritional investigations have provided a strong support to the dramatic ecological evidence represented by the Mediterranean natural experiment. The current momentum towards the impact of Mediterranean diet on longevity has solid biological foundation and does not represent a transient fashion.

References

1. World report on ageing and health. Geneva: World Health Organization; 2015.
2. Global strategy and action plan on ageing and health. Geneva: World Health Organization; 2017. Licence: CC BY-NC-SA3.0 IGO.
3. World Cancer Research Fund/American Institute for Cancer Research: Diet, Nutrition and Physical Activity: a Global Perspective. Continuous Update Project Expert Report; 2018. dietandcancerreport.org.
4. Schulze MB, Martínez-González MA, Fung TT, Lichtenstein AH, Forouhi NG. Food based dietary patterns and chronic disease prevention. BMJ. 2018;13(361):k2396.
5. Bamia C. Dietary patterns in association to cancer incidence and survival: concept, current evidence, and suggestions for future research. Eur J Clin Nutr. 2018;72(6):818–25.
6. Hu FB. Dietary pattern analysis: a new direction in nutritional epidemiology. Curr Opin Lipidol. 2002;13:3–9.
7. Millen BE, Abrams S, Adams-Campbell L, Anderson CA, Brenna JT, Campbell WW, et al. The 2015 dietary guidelines advisory committee scientific report: development and major conclusions. Adv Nutr. 2016;7(3):438–44.
8. Tapsell LC, Neale EP, Satija A, Hu FB. Foods, nutrients, and dietary patterns: interconnections and implications for dietary guidelines. Adv Nutr. 2016;7(3):445–54.

9. Jankovic N, Geelen A, Streppel MT, de Groot LC, Orfanos P, van den Hooven EH, et al. Adherence to a healthy diet according to the World Health Organization guidelines and all-cause mortality in elderly adults from Europe and the United States. Am J Epidemiol. 2014;180(10):978–88.

10. Jankovic N, Geelen A, Winkels RM, Mwungura B, Fedirko V, Jenab M, et al. Consortium on health and ageing: network of cohorts in Europe and the United States (CHANCES). Adherence to the WCRF/AICR dietary recommendations for cancer prevention and risk of cancer in elderly from Europe and the United States: a meta-analysis within the CHANCES project. Cancer Epidemiol Biomarkers Prev. 2017;26(1):136–44.

11. Trichopoulou A, Lagiou P. Healthy traditional Mediterranean diet: an expression of culture, history and lifestyle. Nutr Rev. 1997;55:383–9.

12. Keys A. Seven countries. A multivariate analysis of death and coronary heart disease. Cambridge: Harvard University Press: London; 1980.

13. Trichopoulou A, Costacou T, Bamia C, Trichopoulos D. Adherence to a Mediterranean diet and survival in a Greek population. N Engl J Med. 2003;348 2599–608.

14. Bach-Faig A, Serra-Majem L, Carrasco JL, Roman B, Ngo J, Bertomeu I, et al. The use of indexes evaluating the adherence to the Mediterranean diet in epidemiological studies: a review. Public Health Nutr. 2006;9(1A):132–46.

15. Sofi F, Macchi C, Abbate R, Gensini GF, Casini A. Mediterranean diet and health status: an updated meta-analysis and a proposal for a literature-based adherence score. Public Health Nutr. 2014;17(12):2769–82.

16. Benetou V, Orfanos P, Feskanich D, Michaëlsson K, Pettersson-Kymmer U, Byberg L, et al. Mediterranean diet and hip fracture incidence among older adults: the CHANCES project. Osteoporos Int. 2018;29(7):1591–9.

17. Dinu M, Pagliai G, Casini A, et al. Mediterranean diet and multiple health outcomes: an umbrella review of meta-analyses of observational studies and randomised trials. Eur J Clin Nutr. 2018;72(1):30–43.

18. Eleftheriou D, Benetou V, Trichopoulou A, La Vecchia C, Bamia C. Mediterranean diet and its components in relation to all-cause mortality: meta-analysis. Br J Nutr. 2018;120(10):1081–97.

19. Martinez-Gonzalez MA, Bes-Rastrollo M. Dietary patterns, Mediterranean diet, and cardiovascular disease. Curr Opin Lipidol. 2014;25:20–6.

20. Schwingshackl L, Hoffmann G. Mediterranean dietary pattern, inflammation and endothelial function: a systematic review and meta-analysis of intervention trials. Nutr Metab Cardiovasc Dis. 2014;24:929–39.

21. Gay HC, Rao SG, Vaccarino V, et al. Effects of different dietary interventions on blood pressure: systematic review and meta-analysis of randomized controlled trials. Hypertension. 2016;67:733–9.

22. Ahmad S, Moorthy MV, Demler OV, et al. Assessment of risk factors and biomarkers associated with risk of cardiovascular disease among women consuming a Mediterranean diet. JAMA Netw Open. 2018;1(8):e185708.

23. Estruch R, Ros E, Salas-Salvadó J, Covas MI, Corella D, Arós F, et al.; PREDIMED Study Investigators. Primary prevention of cardiovascular disease with a Mediterranean diet supplemented with extra-virgin olive oil or nuts. N Engl J Med. 2018;378(25):e34.

24. Couto E, Boffetta P, Lagiou P, Ferrari P, Buckland G, Overvad K, et al. Mediterranean dietary pattern and cancer risk in the EPIC cohort. Br J Cancer. 2011;104(9):1493–9.

25. Schwingshackl L, Missbach B, König J, Hoffmann G. Adherence to a Mediterranean diet and risk of diabetes: a systematic review and meta-analysis. Public Health Nutr. 2015;18(7):1292–9.

26. van de Rest O, Berendsen AA, Haveman-Nies A, de Groot LC. Dietary patterns, cognitive decline, and dementia: a systematic review. Adv Nutr. 2015;6(2):154–68.

27. Valls-Pedret C, Sala-Vila A, Serra-Mir M, Corella D, de la Torre R, Martínez-González MÁ, et al. Mediterranean diet and age-related cognitive decline: a randomized clinical trial. JAMA Intern Med. 2015;175(7):1094–103.

28. Trichopoulou A, Bamia C, Trichopoulos D. Anatomy of health effects of Mediterranean diet: Greek EPIC prospective cohort study. BMJ. 2009;23(338):b2337.

29. Diet, physical activity and cardiovascular disease prevention in Europe. European Heart Network; 2011.
30. Sköldstam L, Hagfors L, Johansson G. An experimental study of a Mediterranean diet intervention for patients with rheumatoid arthritis. Ann Rheum Dis. 2003;62(3):208–14.
31. Veronese N, Stubbs B, Noale M, Solmi M, Luchini C, Maggi S. Adherence to the Mediterranean diet is associated with better quality of life: data from the osteoarthritis initiative. Am J Clin Nutr. 2016;104(5):1403–9.
32. Dyer J, Davison G, Marcora SM, Mauger AR. Effect of a Mediterranean type diet on inflammatory and cartilage degradation biomarkers in patients with osteoarthritis. J Nutr Health Aging. 2017;21(5):562–6.
33. Trichopoulou A, Bamia C, Trichopoulos D. Mediterranean diet and survival among patients with coronary heart disease in Greece. Arch Intern Med. 2005;165(8):929–35.
34. de Lorgeril M, Salen P, Martin JL, Monjaud I, Delaye J, Mamelle N. Mediterranean diet, traditional risk factors, and the rate of cardiovascular complications after myocardial infarction: final report of the Lyon Diet Heart Study. Circulation. 1999;99(6):779–85.
35. Frölich S, Lehmann N, Weyers S, Wahl S, Dragano N, Budde T, Kälsch H, et al.; Heinz Nixdorf Recall Study Investigators. Association of dietary patterns with five-year degree and progression of coronary artery calcification in the Heinz Nixdorf Recall Study. Nutr Metab Cardiovasc Dis. 2017;27(11):999–1007.
36. Esposito K, Maiorino MI, Ceriello A, Giugliano D. Prevention and control of type 2 diabetes by Mediterranean diet: a systematic review. Diabetes Res Clin Pract. 2010;89(2):97–102.
37. Ratjen I, Schafmayer C, di Giuseppe R, Waniek S, Plachta-Danielzik S, Koch M, Nöthlings U, Hampe J, Schlesinger S, Lieb W. Postdiagnostic Mediterranean and healthy Nordic dietary patterns are inversely associated with all-cause mortality in long-term colorectal cancer survivors. J Nutr. 2017;147(4):636–44.
38. Trichopoulou A. Diversity v. globalization: traditional foods at the epicentre. Public Health Nutr. 2012;15(6):951–4.
39. Burlingame B, Dernini S. Sustainable diets: the Mediterranean diet an example. Public Health Nutr. 2011;14(12A):2285–7.
40. Martínez-González MÁ, Hershey MS, Zazpe I, Trichopoulou A. Correction: Martínez-González, M.A. et al. Transferability of the Mediterranean diet to non-Mediterranean countries. What is and what is not the Mediterranean diet. Nutrients 2017, 9, 1226. Nutrients. 2018;10(7).

Lifespan and Healthspan Extension by Nutraceuticals: An Overview

11

Sergio Davinelli and Giovanni Scapagnini

11.1 Introduction

Ageing is now an international challenge to healthcare systems in developed countries. It was estimated that, by the year 2050, at least 25% of the population in the developed countries will be older than 65 years, with some regions exceeding 40% [1]. Although the average lifespan has increased over the past two centuries, the number of people living with age-related diseases has also increased [2]. The most common age-related diseases are cancers, cardiovascular diseases, diabetes and Alzheimer's and Parkinson's diseases. The ageing and nutrition research fields have undergone dramatic change in the past two decades. A number of dietary interventions have recently been described to extend the lifespan of multiple model organisms and potentially prevent age-related diseases [3]. Lifespan is currently the most used measure of ageing; however, healthspan, the functional and disease-free period of life, has become the central theme of gerontology and nutrition research [4, 5]. A wide array of studies on several organisms including *Caenorhabditis elegans*, flies, mice and humans have shown that bioactive food compounds such as resveratrol or spermidine may counteract age-related functional decline of cells and tissues and delay the onset of various age-related diseases [6–9]. Bioactive compounds are mainly found in plant-based foods. Observational studies have suggested that foods

S. Davinelli (✉)
Department of Medicine and Health Sciences "V. Tiberio", University of Molise, Campobasso, Italy

Department of Epidemiology, Harvard T. H. Chan School of Public Health, Boston, MA, USA
e-mail: sergio.davinelli@unimol.it

G. Scapagnini
Department of Medicine and Health Sciences "V. Tiberio", University of Molise, Campobasso, Italy
e-mail: giovanni.scapagnini@unimol.it

such as fruits and vegetables, nuts, chocolate and fatty fish, as well as beverages such as tea, wine and coffee, are associated with a wide range of health benefits [10]. Numerous evidence suggests that several plant-based foods may increase cellular resistance to ageing by modulating signalling pathways associated with inflammation, autophagy, and proteotoxic and oxidative stress [11, 12]. More specifically, examples of signalling pathways affected by phytochemicals include upregulation of antioxidant enzymes, sirtuins, DNA repair efficiency and trophic factor production, and inhibition of inflammatory mediators [13–15]. Plant foods contain multiple phytochemicals such as resveratrol (red grapes and peanuts), allicin (garlic), lycopene (tomato), isothiocyanates, indole-3-carbinol, sulphoraphane (cruciferous vegetables), carotenoids (carrots), ellagic acid (pomegranate), myricetin (cranberry), carnosol (rosemary), spermidine (whole grains), genistein and daidzein (beans and soy), catechins and epicatechins (cocoa and green tea). These compounds dissuade herbivorous organisms from consuming plants, and therefore, they are also defined as "noxious' agents. In this context, phytochemicals may counteract ageing by inducing adaptive (hormetic) stress responses in cells. Hormesis is a dose–response phenomenon characterized by a low-dose stimulation and a high-dose inhibition. This mechanism is involved in the biological amplification of adaptive responses and cytoprotective mechanisms. In particular, diets rich in phytochemicals can induce a mild adaptive stress response and modulate hormetic signalling pathways, leading to the improvement in overall cellular functions and performance [16–18]. Although these bioactive compounds may have beneficial physiological effects and represent an emerging strategy through which to promote healthspan, it is crucial to highlight that they are not considered essential macro- or micronutrients. Essential nutrients such as vitamins, minerals and trace elements have established dietary intake recommendations and provide energy and cell materials that regulate metabolism [19]. Recently, various bioactive compounds and/or nutrients associated with health-related effects have been named as nutraceuticals. This category of nutritional agents can include dietary supplements, functional foods and medical foods. The term nutraceutical, a syncretic neologism of the words nutrient and pharmaceutical, was originally coined by Stephen DeFelice, who defined nutraceuticals as 'food or part of a food that provides medical or health benefits, including the prevention and/or treatment of a disease' [20]. Here, we provide an overview of lifespan extension by nutraceuticals in numerous organisms and discuss their potential ability to extend healthspan in humans.

11.2   Nutraceuticals and Lifespan in Model Organisms

It is well established that life expectancy is a modifiable parameter in experimental organisms. Although lifespan extension by nutraceuticals remains under intense investigation, several laboratories have investigated the ability of multiple bioactive constituents from food to extend the lifespan of animal models (Table 11.1) [35]. Resveratrol has been one of the most widely studied molecules in the context of ageing research. Numerous studies have established the ability of resveratrol to

Table 11.1 Lifespan extension by selected nutraceutical compounds in various species

Organism	Compound	Food source	Dose	Mean lifespan extension	Reference
Caenorhabditis elegans	Resveratrol	Grapes and red wine	1 mM	18%	[21]
Drosophila melanogaster	Resveratrol	Grapes and red wine	100 μM	29%	[22]
Saccharomyces cerevisiae	Resveratrol	Grapes and red wine	100 μM	70%	[23]
Apis mellifera	Resveratrol	Grapes and red wine	130 μM	33%	[24]
Nothobranchius furzeri	Resveratrol	Grapes and red wine	2.5 μM	56%	[6]
Nothobranchius guentheri	Resveratrol	Grapes and red wine	100 μL in 200 μg/g food	19%	[25]
Mus musculus (high-calorie diet)	Resveratrol	Grapes and red wine	22.4 ± 0.4 mg kg^{-1} day^{-1}	31%	[26]
Caenorhabditis elegans	Quercetin	Fruits, vegetables, and legumes	70–200 μM	Up to 18%	[27]
Caenorhabditis elegans	Kaempferol	Fruits, vegetables, and legumes	100 μM	13.50%	[28]
Caenorhabditis elegans	Naringenin	Herbs and fruits	100 μM	12.61%	[28]
Caenorhabditis elegans	Myricetin	Vegetables, fruits, nuts and, tea	100 μM	15.08%	[28]
Caenorhabditis elegans	Baicalein	Grains, leafy vegetables, and herbs	100 μM	45%	[29]
Caenorhabditis elegans	Silymarin	Herbs	50 μM	24.8%	[30]
Drosophila melanogaster	Epicatechin	Tea, apples and cocoa	100–8000 μM	Increase	[27]
Drosophila melanogaster	Fisetin	Fruits and vegetables	100 μM	13%	[27]
Drosophila melanogaster	Epigallocatechin	Tea, apples and cocoa	10 mg/mL	Increase (+3.3 days)	[27]
Caenorhabditis elegans	Curcumin	Turmeric (*Curcuma longa*)	20 μM	39%	[31]
Drosophila melanogaster	Curcumin	Turmeric (*Curcuma longa*)	100 μM	19%	[32]
Caenorhabditis elegans	Caffeic acid	Fruits, vegetables, and seeds	300 μM	8%	[33]
Caenorhabditis elegans	Rosmarinic acid	Herbs	200 μM	10%	[33]
Drosophila melanogaster	Spermidine	Soy products, legumes, corn, and grains	1 mM	Up to 30%	[34]
Caenorhabditis elegans	Spermidine	Soy products, legumes, corn, and grains	0.2 mM	Up to 15%	[34]

promote longevity. Resveratrol supplementation increases the mean lifespan of *C. elegans* (flatworm), *Drosophila melanogaster* (fruit fly) and *Saccharomyces cerevisiae* (yeast) by 18%, 29% and 70%, respectively [21–23]. This natural compound also extends the mean lifespan of the common honey bee by up to 33% [24]. In the short-lived fish species *Nothobranchius furzeri* and *Nothobranchius guentheri*, resveratrol extends both median and maximum lifespan and delays age-associated decline of locomotor activity [6, 25, 36]. Additionally, resveratrol administration reduced the risk of death in mice by 31% due to high-calorie feeding and also improved survival in a rat model of hypertension [26, 37]. Flavonoids are polyphenolic compounds and one of the largest groups of phytochemicals found in human diets [38]. The flavonoid quercetin prolonged lifespan in *C. elegans* at concentrations ranging from approximately 70 to 200 μM. Lower or higher concentrations determined no changes or decreased lifespan in these model organisms [27]. These findings suggest that the determination of optimal dose is a crucial aspect to enhance lifespan. Moreover, other flavonoids such as kaempferol, naringenin, myricetin, baicalein and silymarin isolated from a wide variety of fruits, vegetables and tea increased lifespan in *C. elegans* [28–30]. Using the flavonoids epicatechin, fisetin and epigallocatechin gallate found in cocoa powder and green tea extracts, several authors observed an increase in the lifespan of *D. melanogaster* [27]. To date, six mouse studies investigated the influence of flavonoid-containing extracts on lifespan. Two out of six studies testing these extracts observed a lifespan increase in supplemented mice. These flavonoid extracts were a triple combination of blueberry, green tea and pomegranate powder [27]. Curcumin, a polyphenol found in the common spice turmeric, increases the lifespan of yeast, flies and mice. For example, *C. elegans* exposed to 20 μM curcumin lived 39% longer than those supplemented with 200 μM [31]. Curcumin also increases lifespan in *D. melanogaster*, which was accompanied by protection against oxidative stress, improvement in locomotion and chemopreventive effects. However, male flies have greater longevity at a higher concentration than female flies [32]. Several other polyphenolic compounds promote longevity, particularly in *C. elegans*. It was shown that caffeic and rosmarinic acids increased lifespan in worms by 8% and 10%, respectively [33, 39]. More interestingly, the germ of whole grains contains a polyamine, called spermidine, which has been shown to extend lifespan in flies and nematodes. Spermidine is known to inhibit histone acetyltransferases, which results in higher resistance to oxidative stress, to increase autophagy, as well as reduces inflammation and the rates of cell necrosis during ageing [34].

A list of nutraceuticals known to extend lifespan in various species is depicted in Table 11.1.

11.2.1 Mechanisms Associated with Lifespan Extension by Nutraceuticals

We now know that several nutraceuticals may affect ageing-related pathways, revealing potential novel interventions that promote healthy ageing and counteract disease processes [40, 41]. For example, polyphenols may act on AMP-activated protein kinase (AMPK), which is a crucial regulator of metabolic homeostasis and one of the major survival factors in a variety of metabolic stresses [42]. The activity of AMPK declines with age and accelerates the ageing process. The mechanism to explain the deficiency of AMPK activation associated with ageing is under intensive investigation. It seems that there are age-related changes in the function of protein phosphatases, which could be involved in the suppression of AMPK activation with ageing. Additionally, low-grade inflammation, which is present in aged tissues, may be one phenomenon suppressing AMPK signalling. However, AMPK activity can be stimulated by energy deficiency (calorie restriction) and several nutritional agents including many phytochemicals. Currently, it is known that plant-derived compounds stimulate the interaction between AMPK and sterol regulatory element-binding proteins (SREB). In mammals, this mechanism is crucial to activate AMPK and improve healthspan [42]. Additionally, the network of AMPK and its target proteins are associated with so-called longevity factors such as sirtuin 1 (SIRT1), tumour suppressor p53 (p53) and forkhead box protein O (FoxO). AMPK can activate all these enzymes, increasing cellular stress resistance [43, 44]. The signalling pathway associated with mammalian target of rapamycin (mTOR) is one of the best-known pathways leading to autophagy induction. Autophagy permits clearance of misfolded proteins and non-functional organelles that accumulate especially in post-mitotic cells such as those of the nervous system and of skeletal muscle, thus delaying ageing of those tissues. Various polyphenols such as resveratrol, green tea catechins, silibinin, quercetin and curcumin were shown to regulate autophagy through mTOR [45]. Therefore, autophagy can be an important nutritional target to maintain the function of cell and its organelles and recycle defective proteins. Interestingly, many of the above-mentioned pathways are directly or indirectly associated with mitochondrial biogenesis (sirtuins, AMPK) and mitophagy (mTOR, SIRT1), a specialized form of autophagy dedicated to the degradation of damaged mitochondria [46, 47]. A recent study on the effects of urolithins, metabolites of ellagitannins, which are found in the pomegranate fruit, nuts and berries, suggests a crucial role of mitophagy in health and ageing. The authors highlighted the possibility that mitophagy induction by bioactive food compounds may enhance mitochondrial biogenesis and improve healthspan and lifespan [48]. Recent years have seen strong interest in nutraceuticals that can stimulate endogenous antioxidant enzymes (e.g. superoxide dismutase [SOD], catalase [CAT], glutathione peroxidase [GPx], glutathione S-transferase [GST], haeme oxygenase-1 (HO-1), NAD(P)H:quinone acceptor oxidoreductase 1 (NQO1) and other stress resistance systems regulated by nuclear factor erythroid 2-related factor 2 (Nrf2) signalling. Nrf2 controls the basal and induced expression of an array of antioxidant response element–dependent genes. It was well established that bioactive food agents such as resveratrol, curcumin and epicatechins promote cell survival by enhancing the activity of Nrf2 and the expression of its downstream cytoprotective enzymes [49]. The administration of sulphoraphane, which is the main bioactive component of cruciferous vegetables, was associated with increased expression of Nrf2 in different experimental conditions [50]. Also, oleuropein and hydroxytyrosol found in virgin olive oil and extra-virgin olive oil have shown cytoprotective activity against oxidation by activating the Nrf2 pathway [15]. A plethora of studies have demonstrated that there is an interplay between the Nrf2 system and the NF-κB signalling network, which is the principal pathway involved in the modulation of inflammatory responses. The increased activity of NF-κB during ageing is associated with enhanced production of pro-inflammatory cytokines. The role of Nrf2 against inflammation has been related to its ability to antagonize with NF-κB [51, 52]. Several dietary phytochemicals including anthocyanins and other flavonoids found in berry fruits protect cells from pro-oxidant and pro-inflammatory damage through modulation of Nrf2 and inhibition of NF-κB pathways [53].

11.3   Increasing Healthspan by Nutraceuticals: Challenges and Promise

Although nutraceuticals have demonstrated great promise as lifespan extending compounds in model organisms, many important questions remain unanswered on their impact on healthspan in mammals and humans. Indeed, the main issue is to translate findings from model systems to the clinic as well as to the general population. The concept of healthspan, as 'the period of life spent in good health, free from the chronic diseases and disabilities of ageing', is a useful definition that allows us to consider it as a quantitative variable, in the same way that we consider lifespan. However, healthspan is not an easily quantifiable phenotype [5, 54, 55]. To date, there are no metric indices or statistical methods available to measure whether healthspan is extended in a significant manner. Additionally, long-term and large-scale randomized trials in middle-age and elderly subjects are crucial in understanding how nutraceuticals may be involved in healthspan promotion. Unfortunately, it would be very difficult to perform these long-term clinical studies due to enormous cost and both subjects and researcher compliance. Although randomized nutritional trials are critically important to provide more definitive answers on the prevention effects of nutraceuticals, population-based observational studies have suggested that several dietary patterns characterized by consumption of nutraceuticals are associated with healthspan promotion [56, 57].

11.3.1 Nutraceutical Diet and Longevity

Residents of Okinawa, Japan; Loma Linda, California and Ikaria, Greece, have disease-free longevity and often survive past 90 or even 100 years of age. These populations have a preferential attitude towards a plant-based diet. The main characteristics and the health-promoting effects of plant-based diets are depicted in Fig. 11.1.

The traditional Okinawan diet provides about 90% of calories from carbohydrates, but in the vegetable form, they are low in calories but nutritionally dense, particularly with regard to vitamins, minerals and phytochemicals. Okinawans often

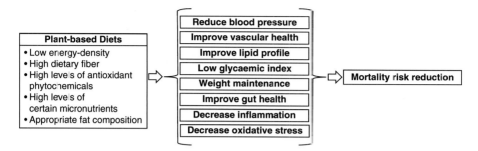

Fig. 11.1 Characteristics and health-promoting effects of plant-based diets

consumed a sweet potato, known as *Ipomoea batatas*, which is a potent food source of free radical quenchers. This tuber is an excellent source of the antioxidant vitamin A (mainly in the form of beta-carotene) and a good source of antioxidant vitamins C and E, and other anti-inflammatory phytochemicals [56]. In Loma Linda, California, there is a community of Seventh Day Adventists. These individuals engage healthy lifestyle practices, and they live longer than the rest of the population. In particular, their vegetarian diet rich in polyphenols is thought to be one of the most likely causes of their extraordinary longevity and well-preserved physiological function. Specific dietary factors that may be involved in their outstanding health include a high intake of fruit, vegetables and nuts. In the most recent Adventist Health Study, a study of 73,308 California Seventh-Day Adventists, vegetarian dietary patterns were associated with lower all-cause mortality and with some reductions in cause-specific mortality [58]. The traditional Mediterranean diet consumed in Southern Europe is characterized by plant foods containing multiple phytochemicals with potential health benefits against cancer and other age-related diseases. Over the past two decades, the Mediterranean diet was not only associated with increased lifespan but also related to a decrease in all-cause mortality. Findings from cross-sectional, longitudinal (observational) and intervention studies in humans support a significant modulatory effect of Mediterranean diet and its bioactive constituents on ageing process. In particular, a meta-analysis of prospective cohort studies showed that a greater adherence to a Mediterranean diet is associated with a significant improvement in health status, as seen by a significant reduction in overall mortality (9%), mortality from cardiovascular diseases (9%), incidence of or mortality from cancer (6%) and incidence of Parkinson's disease and Alzheimer's disease (13%) [59]. It was observed that people in Ikaria Island, Greece, have one of the highest life expectancies in the world. Ikarians are three times more likely to reach the age of 90 years than those in the USA. In this community, long-term adherence to the Mediterranean diet was associated with greater longevity [60]. Moreover, it seems that even an individual bioactive constituent consumed within a diet may be useful in the prevention of age-associated diseases. For example, epidemiological evidence suggests that a diet rich in cocoa polyphenols reduces cardiovascular events in the general population. The Kuna Indians, a population living on the island off the Panama coast, have a diet characterized by high intake of epicatechins from cocoa and a lower blood pressure than other Pan-American populations. The Kuna Indians have low incidence of heart attack and stroke, but after migration to an urban city, they change their diet, and this cardioprotection is lost [61]. Polyphenol-rich foods and/or dietary polyphenol intake, particularly catechins and epicatechins, are inversely associated with chronic diseases such as cardiovascular and neurodegenerative diseases and some cancers. The findings from the Chianti study, a population-based cohort study of older adults living in the Chianti region of Tuscany, Italy, have shown that high total urinary polyphenol concentrations, a nutritional biomarker of polyphenol intake, are associated with reduced all-cause mortality in an elderly, free-living population. The same study, in older adults without dementia, demonstrated that a high intake of these compounds is associated with lower risk of cognitive decline over a 3-year period [62, 63].

11.4 Conclusion

Findings from preclinical investigations as well as observational studies provide evidence of a modulatory effect of diet composition on physiological changes associated with ageing. In general, a high intake of fruits, vegetables, whole grains and nuts may exert health-enhancing effects and promote preservation of physiological function with ageing, enhancing the achievement of longevity. There is a growing list of food bioactive constituents, also known as nutraceuticals, proven to extend lifespan and healthspan in experimental models. Although there is a tremendous amount of work to translate these findings in humans, many nutraceuticals may improve ageing phenotype, modulating fundamental mechanisms such as oxidative stress, inflammation, mitochondrial dysfunction, cellular senescence, impaired proteostasis, autophagy and telomere attrition. Preliminary observational studies involving middle-aged/older humans have reported how bioactive food components may preserve function, compress disability and contribute to enhance healthspan. However, more epidemiological studies and long-term randomized controlled dietary intervention trials with hard clinical endpoints are needed before nutraceuticals can be considered as a feasible option to extend healthspan in humans.

References

1. Harper S. Economic and social implications of aging societies. Science. 2014;346(6209):587–91.
2. Burch JB, Augustine AD, Frieden LA, Hadley E, Howcroft TK, Johnson R, et al. Advances in geroscience: impact on healthspan and chronic disease. J Gerontol A Biol Sci Med Sci. 2014;69(Suppl 1):S1–3.
3. Aiello A, Accardi G, Candore G, Carruba G, Davinelli S, Passarino G, et al. Nutrigerontology: a key for achieving successful ageing and longevity. Immun Ageing. 2016;13:17.
4. Davinelli S, Willcox DC, Scapagnini G. Extending healthy ageing: nutrient sensitive pathway and centenarian population. Immun Ageing. 2012;9:9.
5. Kirkland JL, Peterson C. Healthspan, translation, and new outcomes for animal studies of aging. J Gerontol A Biol Sci Med Sci. 2009;64(2):209–12.

6. Valenzano DR, Terzibasi E, Genade T, Cattaneo A, Domenici L, Cellerino A. Resveratrol prolongs lifespan and retards the onset of age-related markers in a short-lived vertebrate. Curr Biol. 2006;16(3):296–300.
7. Davinelli S, Sapere N, Visentin M, Zella D, Scapagnini G. Enhancement of mitochondrial biogenesis with polyphenols: combined effects of resveratrol and equol in human endothelial cells. Immun Ageing. 2013;10(1):28.
8. Morselli E, Galluzzi L, Kepp O, Criollo A, Maiuri MC, Tavernarakis N, et al. Autophagy mediates pharmacological lifespan extension by spermidine and resveratrol. Aging (Albany NY). 2009;1(12):961–70.
9. Madeo F, Carmona-Gutierrez D, Kepp O, Kroemer G. Spermidine delays aging in humans. Aging (Albany NY). 2018;10(8):2209–11.
10. Mozaffarian D. Dietary and policy priorities for cardiovascular disease, diabetes, and obesity: a comprehensive review. Circulation. 2016;133(2):187–225.
11. Khuda-Bukhsh AR, Das S, Saha SK. Molecular approaches toward targeted cancer prevention with some food plants and their products: inflammatory and other signal pathways. Nutr Cancer. 2014;66(2):194–205.
12. Chattopadhyay D, Thirumurugan K. Longevity promoting efficacies of different plant extracts in lower model organisms. Mech Ageing Dev. 2018;171:47–57.
13. Leonov A, Arlia-Ciommo A, Piano A, Svistkova V, Lutchman V, Medkour Y, et al. Longevity extension by phytochemicals. Molecules. 2015;20(4):6544–72.
14. Corrêa RCG, Peralta RM, Haminiuk CWI, Maciel GM, Bracht A, Ferreira ICFR. New phytochemicals as potential human anti-aging compounds: reality, promise, and challenges. Crit Rev Food Sci Nutr. 2018;58(6):942–57.
15. Davinelli S, Maes M, Corbi G, Zarrelli A, Willcox DC, Scapagnini G. Dietary phytochemicals and neuro-inflammaging: from mechanistic insights to translational challenges. Immun Ageing. 2016;13:16.
16. Son TG, Camandola S, Mattson MP. Hormetic dietary phytochemicals. Neuromolecular Med. 2008;10(4):236–46.
17. Mattson MP. Dietary factors, hormesis and health. Ageing Res Rev. 2008;7(1):43–8.
18. Scapagnini G, Davinelli S, Kaneko T, Koverech G, Koverech A, Calabrese EJ, et al. Dose response biology of resveratrol in obesity. J Cell Commun Signal. 2014;8(4):385–91.
19. Rautiainen S, Sesso HD, Manson JE. Large-scale randomized clinical trials of bioactives and nutrients in relation to human health and disease prevention—lessons from the VITAL and COSMOS trials. Mol Aspects Med. 2018;61:12–7.
20. DeFelice SL. The nutraceutical revolution: its impact on food industry R&D. Trends Food Sci Technol. 1995;6:59–61.
21. Viswanathan M, Kim SK, Berdichevsky A, Guarente L. A role for SIR-2.1 regulation of ER stress response genes in determining C. elegans life span. Dev Cell. 2005;9(5):605–15.
22. Wood JG, Rogina B, Lavu S, Howitz K, Helfand SL, et al. Sirtuin activators mimic caloric restriction and delay ageing in metazoans. Nature. 2004;430(7000):686–9.
23. Howitz KT, Bitterman KJ, Cohen HY, Lamming DW, Lavu S, Wood JG, et al. Small molecule activators of sirtuins extend Saccharomyces cerevisiae lifespan. Nature. 2003;425(6954):191–6.
24. Rascón B, Hubbard BP, Sinclair DA, Amdam GV. The lifespan extension effects of resveratrol are conserved in the honey bee and may be driven by a mechanism related to caloric restriction. Aging (Albany NY). 2012;4(7):499–508.
25. Yu X, Li G. Effects of resveratrol on longevity, cognitive ability and aging-related histological markers in the annual fish Nothobranchius guentheri. Exp Gerontol. 2012;47(12):940–9.
26. Baur JA, Pearson KJ, Price NL, Jamieson HA, Lerin C, Kalra A, et al. Resveratrol improves health and survival of mice on a high-calorie diet. Nature. 2006;444(7117):337–42.
27. Pallauf K, Duckstein N, Rimbach G. A literature review of flavonoids and lifespan in model organisms. Proc Nutr Soc. 2017;76(2):145–62.
28. Grünz G, Haas K, Soukup S, Klingenspor M, Kulling SE, Daniel H, et al. Structural features and bioavailability of four flavonoids and their implications for lifespan-extending and antioxidant actions in C. elegans. Mech Ageing Dev. 2012;133(1):1–10.

29. Havermann S, Rohrig R, Chovolou Y, Humpf HU, Wätjen W. Molecular effects of baicalein in Hct116 cells and *Caenorhabditis elegans*: activation of the Nrf2 signaling pathway and prolongation of lifespan. J Agric Food Chem. 2013;61(9):2158–64.
30. Kumar J, Park KC, Awasthi A, Prasad B. Silymarin extends lifespan and reduces proteotoxicity in *C. elegans* Alzheimer's model. CNS Neurol Disord Drug Targets. 2015;14(2):295–302.
31. Liao VH, Yu CW, Chu YJ, Li WH, Hsieh YC, Wang TT. Curcumin-mediated lifespan extension in *Caenorhabditis elegans*. Mech Ageing Dev. 2011;132(10):480–7.
32. Lee KS, Lee BS, Semnani S, Avanesian A, Um CY, Jeon HJ, et al. Curcumin extends life span, improves health span, and modulates the expression of age-associated aging genes in *Drosophila melanogaster*. Rejuvenation Res. 2010;13(5):561–70.
33. Pietsch K, Saul N, Chakrabarti S, Stürzenbaum SR, Menzel R, Steinberg CE. Hormetins, anti-oxidants and prooxidants: defining quercetin-, caffeic acid- and rosmarinic acid-mediated life extension in *C. elegans*. Biogerontology. 2011;12(4):329–47.
34. Eisenberg T, Knauer H, Schauer A, Büttner S, Ruckenstuhl C, Carmona-Gutierrez D, et al. Nat Cell Biol. 2009;11(11):1305–14.
35. Fontana L, Partridge L. Promoting health and longevity through diet: from model organisms to humans. Cell. 2015;161(1):106–18.
36. Genade T, Lang DM. Resveratrol extends lifespan and preserves glia but not neurons of the *Nothobranchius guentheri* optic tectum. Exp Gerontol. 2013;48(2):202–12.
37. Rimbaud S, Ruiz M, Piquereau J, Mateo P, Fortin D, Veksler V, et al. Resveratrol improves survival, hemodynamics and energetics in a rat model of hypertension leading to heart failure. PLoS One. 2011;6(10):e26391.
38. Chun OK, Chung SJ, Song WO. Estimated dietary flavonoid intake and major food sources of U.S. adults. J Nutr. 2007;137(5):1244–52.
39. Uysal U, Seremet S, Lamping JW, Adams JM, Liu DY, Swerdlow RH, et al. Consumption of polyphenol plants may slow aging and associated diseases. Curr Pharm Des. 2013;19(34):6094–111.
40. Davinelli S, Calabrese V, Zella D, Scapagnini G. Epigenetic nutraceutical diets in Alzheimer's disease. J Nutr Health Aging. 2014;18(9):800–5.
41. Lee J, Jo DG, Park D, Chung HY, Mattson MP. Adaptive cellular stress pathways as thera-peutic targets of dietary phytochemicals: focus on the nervous system. Pharmacol Rev. 2014;66(3):815–68.
42. Li Y, Xu S, Mihaylova MM, Zheng B, Hou X, Jiang B, et al. AMPK phosphorylates and inhib-its SREBP activity to attenuate hepatic steatosis and atherosclerosis in diet-induced insulin-resistant mice. Cell Metab. 2011;13(4):376–88.
43. Cantó C, Gerhart-Hines Z, Feige JN, Lagouge M, Noriega L, Milne JC, et al. AMPK reg-ulates energy expenditure by modulating NAD+ metabolism and SIRT1 activity. Nature. 2009;458(7241):1056–60.
44. Salminen A, Kaarniranta K. AMP-activated protein kinase (AMPK) controls the aging process via an integrated signaling network. Ageing Res Rev. 2012;11(2):230–41.
45. Pallauf K, Rimbach G. Autophagy, polyphenols and healthy ageing. Ageing Res Rev. 2013;12(1):237–52.
46. López-Otín C, Blasco MA, Partridge L, Serrano M, Kroemer G. The hallmarks of aging. Cell. 2013;153(6):1194–217.
47. Fang EF, Scheibye-Knudsen M, Chua KF, Mattson MP, Croteau DL, Bohr VA. Nuclear DNA damage signalling to mitochondria in ageing. Nat Rev Mol Cell Biol. 2016;17(5):308–21.
48. Ryu D, Mouchiroud L, Andreux PA, Katsyuba E, Moullan N, Nicolet-Dit-Félix AA, et al. Urolithin A induces mitophagy and prolongs lifespan in *C. elegans* and increases muscle func-tion in rodents. Nat Med. 2016;22(8):879–88.
49. Scapagnini G, Vasto S, Abraham NG, Caruso C, Zella D, Fabio G. Modulation of Nrf2/ARE pathway by food polyphenols: a nutritional neuroprotective strategy for cognitive and neuro-degenerative disorders. Mol Neurobiol. 2011;44(2):192–201.

50. Briones-Herrera A, Eugenio-Pérez D, Reyes-Ocampo JG, Rivera-Mancía S, Pedraza-Chaverri J. New highlights on the health-improving effects of sulforaphane. Food Funct. 2018;9(5):2589–606.
51. Nair S, Doh ST, Chan JY, Kong AN, Cai L. Regulatory potential for concerted modulation of Nrf2- and Nfkb1-mediated gene expression in inflammation and carcinogenesis. Br J Cancer. 2008;99(12):2070–82.
52. Li W, Khor TO, Xu C, Shen G, Jeong WS, Yu S, et al. Activation of Nrf2-antioxidant signaling attenuates NFkappaB-inflammatory response and elicits apoptosis. Biochem Pharmacol. 2008;76(11):1485–9.
53. de Pascual-Teresa S. Molecular mechanisms involved in the cardiovascular and neuroprotective effects of anthocyanins. Arch Biochem Biophys. 2014;559:68–74.
54. Richardson A, Fischer KE, Speakman JR, de Cabo R, Mitchell SJ, Peterson CA, et al. Measures of healthspan as indices of aging in mice-A recommendation. J Gerontol A Biol Sci Med Sci. 2016;71(4):427–30.
55. Tatar M. Can we develop genetically tractable models to assess healthspan (rather than life span) in animal models? J Gerontol A Biol Sci Med Sci. 2009;64(2):161–3.
56. Willcox DC, Scapagnini G, Willcox BJ. Healthy aging diets other than the Mediterranean: a focus on the Okinawan diet. Mech Ageing Dev. 2014;136–137:148–62.
57. Davinelli S, Trichopoulou A, Corbi G, De Vivo I, Scapagnini G. The potential nutrigeroprotective role of Mediterranean diet and its functional components on telomere length dynamics. Ageing Res Rev. 2019;49:1–10. https://doi.org/10.1016/j.arr.2018.11.001.
58. Orlich MJ, Singh PN, Sabaté J, Jaceldo-Siegl K, Fan J, Knutsen S, et al. Vegetarian dietary patterns and mortality in Adventist Health Study 2. JAMA Intern Med. 2013;173(13):1230–8.
59. Sofi F, Cesari F, Abbate R, Gensini GF, Casini A. Adherence to Mediterranean diet and health status: meta-analysis. BMJ. 2008;337:a1344.
60. Stefanadis CI. Unveiling the secrets of longevity: the Ikaria study. Hellenic J Cardiol. 2011;52(5):479–80.
61. Hollenberg NK, Martinez G, McCullough M, Meinking T, Passan D, Preston M, et al. Aging, acculturation, salt intake, and hypertension in the Kuna of Panama. Hypertension. 1997;29(1 Pt 2):171–6.
62. Zamora-Ros R, Rabassa M, Cherubini A, Urpí-Sardà M, Bandinelli S, Ferrucci L, et al. High concentrations of a urinary biomarker of polyphenol intake are associated with decreased mortality in older adults. J Nutr. 2013;143(9):1445–50.
63. Rabassa M, Cherubini A, Zamora-Ros R, Urpi-Sarda M, Bandinelli S, Ferrucci L, et al. Low levels of a urinary biomarker of dietary polyphenol are associated with substantial cognitive decline over a 3-year period in older adults: the Invecchiare in Chianti Study. J Am Geriatr Soc. 2015;63(5):938–46.

Printed in the United States
By Bookmasters